Mobility and Globalization in the Aftermath of COVID-19

"The book is a vivid testimony of how this pandemic chained and changed us and created new geographies of exclusion. Dr. Korstanje and George re-examine the notion of hospitality in the West in times of crisis and re-open the debate about the emerging conflictual relationship with the non-Western Other, whom they call the "undesired guest". Dr. Korstanje and George's work is indeed an important addition to the literature; it continues the unfinished debate on colonialism and paves the way for a new one on the end of hospitality in the West in times of COVID-19."

—Bootheina Majoul, *Carthage University, Tunisia*

"This is a fascinating and timely examination of hospitality and colonialism. Korstanje and George draw widely across social sciences and historical literature to develop a postcolonial and post-COVID-19 understanding of where travel, tourism and hospitality might take us next."

—Jonathan Skinner, *University of Roehampton, UK*

Maximiliano E. Korstanje • Babu George

Mobility and Globalization in the Aftermath of COVID-19

Emerging New Geographies in a Locked World

palgrave
macmillan

Maximiliano E. Korstanje
Department of Economics
University of Palermo
Buenos Aires, Argentina

Babu George
School of Business
Christian Brothers University
Memphis, TN, USA

ISBN 978-3-030-78844-5 ISBN 978-3-030-78845-2 (eBook)
https://doi.org/10.1007/978-3-030-78845-2

This Palgrave Macmillan imprint is published by the registered company Springer Nature Switzerland AG.
The registered company address is: Gewerbestrasse 11, 6330 Cham, Switzerland

Foreword 1: COVID, Passports, and Terror

I want to continue several tropes from Korstanje and George's book. It correctly points to a new global era of terror beginning with 9/11—that is, the attacks on the World Trade Center and Pentagon in the United States on September 11, 2001. COVID-19 is a continuation of the same regime of terror initiated 20 years ago. Not coincidentally the current era of terror coincides with a crisis in the world system of capital as described by Immanuel Wallerstein.[1] The Global War on Terror announced by US president George W. Bush was the first step in trying to contain the effects of the capitalist crisis. COVID or more accurately governments' putative attempts to control the disease are the second step. At this point in 2021, perhaps it is not too early to identify the Great Reset[2] as the third and final step to stem the breakdown of the world system. All three steps should be seen as tools used by the global ruling class to maintain their position of hegemonic control.

As Korstanje and George point out, tourism is terrorism by other means. European imperialism kicked off the capitalist system. Whether fifteenth-century conquest, nineteenth-century colonialism, or twentieth-century neoliberalism, the objective is always the same: exploitation and extraction of wealth under forced conditions. In case the natives get restless, there are Gatling guns and mercenaries close at hand, while terror and the so-called wars on it are useful for the periodic invasions of the non-West (here translated as the territories to be exploited and controlled). Once pacified, the

[1] *World-System Analysis: An Introduction* (Durham, NC: Duke University Press, 2004).

[2] Klaus Schwab and Thierry Mallert, *COVID-19: The Great Reset* (Geneva, SZ: Forum Publishing, 2021).

colonies had to be subject to the diversions of the equally captive, but no less exploited, working classes of the metropoles—hence tourism and passports. In the last 30 years, marked by the collapse of the USSR in 1990–1991, the United States has taken over the role of enforcer for the global ruling class with military bases and naval task forces, now augmented by cyber armies and "Space Forces," spread world over. Travel for global cops and soldiers is free, while everyone else needs a "passport" soon to be augmented by a COVID passport that will allow the pure to easily pass into and through the bodies of the exploited billions in the world. The exploited are not allowed antibodies of resistance to penetration. Any signs of resistance results in shock and awe in Afghanistan, Iraq, Syria, or Yemen. T cells and antibodies are enveloped and destroyed by the invaders.

Controlling populations through terror is far from new. Ancient empires used it in two forms: physical violence (slaughter, torture, and similar variants) coupled with institutionalized superstitions, alternatively known as state religions. Louis Althusser used different terms: state repressive apparatuses and state ideological apparatuses. Althusser argued that their purpose and employment ensured reproduction of the social order so that those on top stayed there.[3] To understand the current regimes of fear and terror, all that is necessary is to observe the source of violence and how it is used, and equally observe the various campaigns of propaganda and public relations, who is producing and promulgating them, and how they are doing so. The overwhelming amount of both repression and ideology emanate from the state apparatuses of the ruling class of global capital, led by the United States and certain so-called non-governmental bodies such as the World Health Organization, Human Rights Watch, and Amnesty International.

These efforts by the global ruling class have an unfortunate consequence for most of humanity. Not only do they keep people from rising up against their rulers but they divert attention to wholly manufactured mass terrors. In the twenty-first century, humanity faces several real threats, but the terror regimes preclude lucid analyses and informed, considered responses to ameliorate the threats. One is drawn back to Marx's notional Hegelian reference in *The Eighteenth Brumaire of Louis Bonaparte*: history repeats itself, the first time as tragedy, the second as farce. In the second decade of this century we are still in the tragedy phase.

Emeritus Professor, Buffalo State College Geoffrey R. Skoll

[3] Louis Althusser, *On the Reproduction of Capitalism*. Translated by G M Goshgarian (New York: Verso, 2014 (1995)).

Foreword 2: The Grand Experiment

We have been involved in a massive natural experiment for surviving a global pandemic: will scientific understanding about the disease guide political and public behavior? In the United States, public health measures about COVID-19 based on sound science were routinely rejected out of fear and for political reasons. And lots of people died because of this. Mistrust fueled by the politics of fear often trumps knowledge and science. It was easier when we didn't have the choice about who to save and how to do it.

The rationale and implementation of some of these choices remain grounded in prior assumptions about who is worthy, preferred, legitimate, eligible, and strategically most viable to serve our interests. Discussions about the pandemic in the United States are taking place at the same time that immigrants at the Mexican border are sifted through eligibility criteria. In other words, choices involve me/my/us and outsiders/others. Many centuries old mantras about differences persist to regulate access and mobility. This is why colonialism and justifications about humanity, morality, and racism matter. *Mobility and Globalization in the Aftermath of COVID-19* captures the conflict between risk reduction through science, medicine, and policy, on the one hand, and equality and tolerance, on the other hand. The ensuing paradox of the COVID-19 pandemic is, how do you stop it without exacerbating the inequities and power differences that contributed to its emergence? On the surface, we are all helping produce a global economic order, but that order maintains massive social, political, and economic imbalances because profit entails differential participation in production, trade, consumption, migration, and travel.

Maximiliano Korstanje and Babu Geoege appreciate the contradictions of globalization that grease inequalities through capitalism's rationale trade, consumption, travel, and regulatory practices. So did Donald Trump. During 2015 and 2021 he told more than 30,000 lies in some 25,000 tweets promoting the propaganda of fear about threats to Americans by immigrants, Muslims, outsiders. He railed against enemies, which encompassed Democrats, the news media, many allies engaged in treaties, and regulations that promoted more global integration and cooperation. All were outsiders/others bent on destruction of the American Way of Life. Nazi sympathizers, white supremacists, and avowed racists were okay with him as long as they put "America first."

Nationalism was heightened with the pandemic, which Trump denied was real, then lied about, and criticized science and public health officials for urging changes. According to Deborah Birx, Trump's coronavirus advisor, sound public health policy would have saved hundreds of thousands of lives. But those actions entailed restricting public behavior and business, which would have been bad economically and politically for the Trump administration. So, he refused, instead assuring Americans that the virus would go away, and he blamed China, referring to the virus as Kung Fu flu. These derogatory terms sparked an increase in hate crimes and attacks against Asian Americans. As more people died, many followers denied the virus was real, refused to wear masks, openly defied public health orders to not meet in groups, and attacked public officials who supported the mandates. Even members of Congress agreed with him about the virus as well as his false claim that he did not lose the 2020 presidential election but it was rigged. And they followed his urging to attack the Capitol on January 6, 2021, in a battle that led to the deaths of police officers.

The paradox of control over COVID 19 did not end with Trump. More people became actual or potential outsiders/others as the vaccine became available. The vaccines—deemed to be effective—had to be regulated, produced, shipped, administered, and injected into people's arms in many countries that were differentiated by age, religion, race, ethnicity, gender, and income. While people in other countries lacked access to vaccines, many people in the United States, including millions of white Evangelicals, refused vaccines because of religious dogma and past governmental lies and abuses. The movement toward a post-pandemic order would continue to be complicated by rules and restrictions on travel about those who had been tested and vaccinated and those who were not. Covid passports for Americans to travel were considered as another level of acceptable membership and inclusion. But the governor of Florida, who threatened to file a lawsuit to permit cruise ships to restart tourist activities, banned Covid passports in the state. Contradictions are consistent when pandering to the pandemic. As Korstanje and Babu George note in the concluding chapter, "Doubtless, the origins of modern terrorism and tourism have the same roots."

Regents' Professor Emeritus
Arizona State University

David L. Altheide

Acknowledgment

We want to thank Palgrave Macmillan staff Anne-Kathrin Birchley-Brun and Charanya Manoharan for all their support. We also thank professors David Altheide and Geoffrey Skoll for their positive praise, as well as our families for providing us the time this project demanded.

Contents

CHAPTER 1

Mobility in and of the Empire: A Colonialist History of Mobility

Introduction

Although colonialism links mainly to the European expansion that originated in the fifteenth century, to be exact with the discovery of the Americas, it is no less true that there are many sides to colonialism. In this vein, historians agree with the fact that colonialism can be defined as a policy designed to retain and exercise authority over external (overseas) territories while imposing the proper language, culture, and laws upon natives. These territories are systematically administered by a colonial head (officer) who endorses his authority to the metropolitan elite (Hoffman, 2017). The colonial office not only embodies the authority of the monarch but centralizes the interpretation of the law. These two combined factors very well cement the power of the colonial rule over the locals. To some extent, colonialism exhibits a practice of political domination as well as a cultural (sometimes troublesome) encounter between hosts and guests (Smith, 2012). It is important to note that some voices do not distinguish colonialism from imperialism, using both terms to refer to the same object of study alike. Etymologically speaking, the word *colony* stems from the Latin *colonus*, which means *settler* or *farmer*. Imperialism, which comes from the Latin *imperium* (commanding), evokes the lines of authorities of one territory or administration over others. While the former signals a forced migratory movement in quest of new lands to settle, the latter

M. E. Korstanje, B. George, *Mobility and Globalization in the Aftermath of COVID-19*,
https://doi.org/10.1007/978-3-030-78845-2_1

refers to a political alliance (Cowen & Shenton, 1991; Horvath, 1972; Hill, 2014). At a closer look, economic exploitation is a necessary but not the sufficient condition for colonialism. As Ania Loomba (2015) puts it, the problem of economic—if not political—exploitation traverses the field of cultural studies to explain the genesis of colonialism. To be more precise, the idea of exploitation is misjudged by historians as a passive *subordination*. Under some contexts, the local ruling elite empowered the colonial officers to impose a regime that otherwise would be rejected by the neighboring kinship. Like the case of the Spanish conquest of the Americas, the opposite is equally true; some aboriginal groups that Aztecs enslaved and the Incas embraced the Spanish conquerors enthusiastically as real emancipators (Calvo, 1996). Beyond the disputes of colonialism or postcolonialism, the intersection of literary and colonial expansion is still clear (Nodelman, 1992; Mishra & Hodge, 1991). Literary studies debated in some detail the inflection of colonial travels, which generated narratives and discourses oriented toward molding and controlling the "non-Western native". The technological revolution, which accelerated such an expansion, created specific colonial subjectivities (bodies) finely ingrained in stories that travelers wrote during their expeditions (Pratt, 2007). The history of colonialism is a history eloquently written by a process marked by a cultural hybridization (Ashcroft et al., 2003).

It is safe to say that the mission of writing a book about colonialism and travels is not an easy task. It appears to be a daunting and challenging mission especially because of the over-abundance of publications and works in the field. Secondly and most importantly, the term *colonialism* has received scholarly attention in recent years, but the concept obscures more than it clarifies. Still further, the concept has some limitations to be operationalized for empirical research (Blauner, 1969; Horvath, 1972; Cooper, 2005). The reasons that led colonial European powers to expand their hegemonies beyond their borders, not only indexing new economies but also disposing of military occupation in some peripheral areas, have not been explained with clarity, at least to date (Lehning, 2013). Those theories that take colonialism as an object of study can be grouped into two clear-cut poles. Some studies frame European colonialism as the need to expand the internal trade beyond the national borders to create an economic center-periphery dependency (Hobsbawm, 1954; King, 2015; Bandarage, 2019). This academic wave should be distinguished from those works oriented toward discussing the political implications cemented by a centralized authority embraced by the nation-state (Bassin, 1987;

Gellner, 1998; Pollock, 2004). Academicians acknowledge that there is a clear and manifest misunderstanding revolving around the commonalities and differences between colonialism and civilization. In his seminal book *A Study of History*, Arnold Toynbee (1951) explains that, far from being static, the history of civilization tells us of societies which are shaped and placed in dialogue with dominant empires to boost the necessary technological breakthrough for ensuring the expansion. This privileged elite is originally followed by an internal uncreative proletariat as well as an external (complacent) proletariat which is hired to protect the city when the empire is in jeopardy. This moot point will be developed in this introductory chapter and in the successive chapters of the book.

The book's main aim is not to focus on the history and evolution of European colonialism but to understand how travel and travel behavior have changed from colonialism to the recent virus outbreak in Wuhan (China), known as COVID-19. To be more exact, the book looks to describe how the Western conceptions of the "Otherness" and hospitality have radically shifted (especially after the terrorist attacks on the US soil in 2001). At first glance, travels look more individualist and solipsist today than in past decades. Needless to say that the attacks on the World Trade Center in September 2001, a foundational event associated with other global threats such as COVID-19, has under mined the trust in the "alterity", closing the doors of heaven (Korstanje, 2018). To put things in perspective, the present book intends to debate to what extent Western hospitality is dying, laying the foundations toward a new radical understanding of the phenomenon. The current socio economic crisis ignited by the outbreak of COVID-19 paralyzed the global trade and stopped the world. However, far from being a foundational event, as some scholars believe, the event reaffirms a previous tendency that originated in 2001: the end of hospitality, as least as we know it. Basically, in the post-9/11 context, a radical fear of strangers is culturally imposed as a new form of geopolitics. Terrorism allows policies and programs that otherwise would be widely rejected. This sentiment of fear not only accelerates a social decomposition but closes the doors of the West to the "non-Western Other". The global discourse of fear changes the external geopolitical relations among states while it internally undermines critical thinking (Altheide, 2006, 2007). In the post-9/11 days, the foreigner is demonized as an undesired guest (Korstanje, 2018). Now, the war against terror sets the pace for the war against an invisible virus. Are COVID-19 and terrorism similar issues?

COVID-19 has brought the global mobilities to a grinding halt, affecting negatively the most powerful economies, and altered radically the travel behavior, shifting the geopolitical relations and territorial geographies. Although terrorism and COVID-19 have different origins, both—and the resulting discourses—have commonalities which merit inspection. The "non-Western Other" seems to be suspected and feared as a potential carrier (terrorist) of the virus. Whereas in the post-9/11 contexts the "Other" was seen as a potential terrorist who looked like us, today all of us are potential terrorists (carriers) who affect the public order. The imposition of strict lockdowns—associated with the closure of borders, airspace, and free circulation, even international flights and travels—is a clear mark of a much deeper social decomposition that originated in 2001: the fear of strangers, the cultural touchstone of Western culture. The coronavirus is gradually altering our social relations as well as the morphologies of travels and borders. Old and long-dormant disputes, among provinces or states, are revived because of the pandemic. Within the same nation-states, counties, provinces, regions, and even cities adopt different health protocols and travel requirements for outsider tourists. In some cases, conational tourists are simply rejected because they do not meet requirements like round-trip ticket, COVID-19 antigen tests, or health travel insurance. The migratory requirements historically expected of international tourists are being asked for domestic tourists. This happens mainly because the world is gradually experiencing a feudalization process as never before. As Korstanje (2020) puts it, John Urry coined the term *the tourist gaze* to denote an emerging ethos where gazing new cultures, people, and landscapes is associated with the needs of possessing "the Other". What finally visitors gaze coincides with a cultural matrix that marks what is visually desired or not. COVID-19 has turned this world upside down. Foreign tourists are demonized and seen as "potential threats" which should be contained and locked down. As least as Urry imagined it, the tourist gaze concept should be revisited according to what Korstanje coins the "wicked gaze". This means that the tourist becomes an *undesired guest* (Korstanje, 2020). The durable effects of COVID-19 lead us to interrogate further the nature and evolution of the liberal nation-state and the constructions of the "Other". Knowledge production and science have played a leading role in the configuration of travels, expeditions, and the invention of the alterity. In the next section, we return for a moment to the legacy of Toynbee to elucidate the differences between civilizations and empires.

Civilizations, Rise and the Fall of Giants

Over centuries, philosophers and historians have devoted efforts to trace back the socio-economic backgrounds of civilizations. From Cicero to Hobbes, different academic figures have inspected the nature and expansion of civilizations (Turchin, 2007). While Gibbon (1887) elucidates the rise and collapse of the Roman Empire, Charles Tilly (2018) and Noah Eisenstadt (1967) give some reflections that help to understand the conditions of expansion and decline of civilizations. Although the literature which explores civilization history abounds, there is no encyclopedic compendium (edition) like *A Study of History*, the seminal collection authored by Arnold Joseph Toynbee. Without any doubt, Toynbee is the world's most-read and most-cited scholar. His books have been translated into numerous languages and his legacy continues being fruitful for us today. Having said this, he starts from the premise that the figure of history is never static; rather it can be dissected because of what he dubbed as *mimesis*, which means an imitation process where each generation or civilization mimicries the older one. Scholars mistakenly emphasize the need to differentiate between the history of sophisticated civilizations and primitive societies. Per Toynbee, even if the concept of mimesis is still present in all human organization regardless of its complexity or cultural background, it takes different paths. He goes on to eloquently write:

> An essential difference between civilizations and primitive societies as we know them (the caveat will be found to be important) as the direction taken by mimesis or imitation. Mimesis is a generic feature of all social life. Its operation can be observed both in primitive societies and in civilizations, in every social activity from the imitation of the style of film-arts by their humbler sisters upwards. It operates, however, in a different direction in the two species of society. In primitive societies, as we know them, mimesis is directed towards the older generation and towards dead ancestors who stand, unseen but not unfelt, at the back of the living elders … on the other hand, in societies in process of civilization, mimesis is directed towards creative personalities who command a following because they are pioneers. (Toynbee, 1951: 49)

As the above-cited excerpt shows, Toynbee holds that a dominant minority, which lays the foundations toward a creative (technical) revolution, marks the pace for the consolidation of a civilization. His book offers countless examples and descriptions of different (civilizations) empires. His method, known as *shared history*, provides fresh insight into the evolution and the collapse of structures that range from Egyptian Hellenic society to Minoan, Sinic, and Andean civilizations. He acknowledges that historians should work on mechanical laws which explain the evolution of social facts. Unlike the study of primitive groups studied by ethnographers, so to speak in the site, historians struggle to decipher a past that has not always taken place as it is imagined. This represents one of the challenges of historians. Neither race nor the geographical location explains with accuracy why empires rise and fall. The technological breakthrough, which is started by the dominant minority, opens the doors to index new economies and territories, but (as a Trojan Horse) it erodes the civilization from within. The petrifaction of creativity to deal with outsiders and the environment and internal conflicts with other pressure groups explain the loss of command of the ruling elite. As the civilization grows and expands its institutions and legitimacy of the leaders decrease. For Toynbee, this lack of trust in the leaders paves the way for internal disputes, leading to fragmentation and decline. Here we find a parallel between Toynbee's and Max Weber's arguments. Both say overtly that the original leading groups keep their legitimacy through charisma, but once they die, the followers should adopt technocratic or bureaucratic forms of control oriented toward dominating the "Other". As Toynbee opines,

> When a civilization is in decline it sometimes happens that a particular technique, that has been both feasible and profitable during the growth-stage, now begins to encounter social obstacles and to yield diminishing economic returns; if it becomes patently unremunerative it may be deliberately abandoned. In such a case it would be obviously a complete inversion of the true order of cause and effect to suggest that the abandonment of the technique in such a circumstances was due to a technical inability to practice it and that this technical inability was a cause of the breakdown of the civilization. (Ibid: 255)

Here is precisely where the problem of *mimesis* lies. The risk of collapse is around the corner simply because the faculty of mimesis, which is finely rooted in history, ushers mankind to a much deeper mechanization process that regulates human relations. By mediating between citizens and their institutions, mimesis plays an active role in addressing and dealing with external stimuli. When the efficiency to cope with these external challenges is affected, the ruling elite deploys some alternative instruments to demonize or seduce the outsiders (external proletariat). Sometimes the intelligentsia—formed by renowned scholars—plays an active role in drawing an idealized history of the civilization to keep the social order, but sooner or later, the dominant elite loses control when outsiders press to enter. This happens simply because when the empire seems to be growing, its culture permeates to neighboring groups covering vast regions, but when it breaks down, this charm ceases to directly affect the strangers. Considered as the war band of *barbarians*, outsiders exert pressure to take control of the society. In this vein, Toynbee uses the Greek term *palingenesia* to denote the cycle of the disintegration of an empire which gradually is adopted—and continued—by another new empire. This process is not limited to a dead point; it traverses to a creative new stage where the oppressed external or internal proletariats embrace a new rebirth (renovation).

As previously mentioned, Toynbee's contributions to our argument correspond in two clear-cut directions. On the one hand, the figure of the (outcast) "Other" occupies a central position in the social imaginary of empires, a point we shall discuss critically in the following chapters. On the other, the sense of discovery seems to be vital during the imperial expansion European Colonialism, western knowledge and travels.

European Colonialism, Knowledge, and Travels

It is tempting to admit that knowledge, colonialism, and travels were inextricably intertwined. As James Clifford notably observes, social sciences have overlooked the significance of travels in the process of knowledge production. The term *theory*, which derives from the Greek word *theorein*, describes a complex process of physical movement (travels) to attend events celebrated in other cities. On their return, visitors were obliged to tell the king their experiences. Clifford (1989, 1997) energetically emphasizes on the act of traveling as a foundational event that recreates durable knowledge. One of the trailblazing books in the genre of literary studies is authored by Marie Louise Pratt (*Imperial Eyes*). In this book, she carefully reviews the roots of imperial discourses, laying the foundations for a better understanding of the role played by literature in general and more specifically travel writing in the formation of the "non-Western Other". European visitors imposed a specific sense of identity that was framed by the mainstream cultural values of Western civilization. Emanating from the periphery's economic core, the colonial ideology dissociated the local native voices from the European lord. As Pratt explains, imperial domination is never applied in the same way over locals. In fact, dominators take the so-called superiority for granted. European colonialism was based on two combinable mechanisms. On the one hand, Pratt speaks of scientific knowledge and the needs of finding a classificatory system to understand the external world. The first expeditions of Karl Linneo, a botanist, urged the development of a classificatory system. In this way, travels and expeditions shaped a global consciousness which altered the old borders of Europe. On the other, travel literature (writings) sheds light on the habits, customs, and even the aborigines' political organization. These texts served not only as a form of cultural entertainment for a vast readership but as an ideologized discourse that framed gradually a center-periphery dependency. What is equally important, the so-called superiority of Europeans took place instead of the figure of the *noble savage*, where the native was portrayed as an innocent, child-like "Other" who needed to be re-educated according to the European values. This romantic discourse was nourished by countless personal stories, experiences, and novels characterized by the urgency of *being there* to understand the alterity. With the benefit of hindsight, travelers were painted as almighty agents who gazed everything at its pace. Pratt coins the term *imperial eye* to note *a gaze* that interrogates further the native through the articulation of imagined landscapes. The native lacks history, name, or identity and is subject to the traveler because the traveler's reasoning is never questioned. When European lords mark the "alterity"—probably narrating their lives—they avoid being marked. If colonial expeditions were historically motivated by the quest for scientific answers, it is no less true that the resulted documents drew the borders of the modern nation-state. Fieldworkers were educated to write about everything they inspected, paving the way for creating a new system of knowledge that was finally made from geographies, rivers, roads, and customs, a type of inventory for the imperial order. The scientific knowledge was the maiden of colonial expansion (Pratt, 2007). This is exactly what Marvin Harris dubs as *the rise of anthropological theory* when he discusses the complex relations between modern anthropology—ethnography—and the colonial rule. The first ethnographers interested in protecting the local culture produced a robust knowledge that the colonial officers read and used to control the natives. The point authorizes Harris to recognize that colonialism and anthropology were inevitably entwined. The notion of ***being there***, which was discussed earlier, linked to a second element vital for the colonial expansion: the archetype of the ***noble savage***. The fact is that anthropology evolved around an old paradigm punctuating that pre-modern Europe disappeared because of the advance of industrialism. The old feudal institutions which characterized the daily life in medieval times were replaced by a technological revolution ignited by modernism in the eighteenth and nineteenth centuries. In the same way, indigenous cultures run the same risk of exhaustion. The idea of protection was invariably associated with colonial paternalism and situated Europeans in the moral duty to protect these noble savages (Harris, 2001). The archetype of the noble savage, so to speak, offers a personified character epitomizing the notion of the indigene, as an outsider wild "Other" who the civilization process has not corrupted. In consequence, he was symbolized as a child-like or innocent person who must be educated to be protected from the vices of the industrial world. This romantic primitivism accompanied Western civilization's philosophy from its outset (Fryd, 1995; Pagden, 1982; Harris, 2001). Echoing Edward Said, knowledge production is based on the old colonial practices of intervention that claimed so-called objectivity regarding the primitive man and modified their habits, cultures, and customs once they were being studied. From the eye of history, all human disciplines claimed a sense of objectivity that turned to be inadequate to describe the natives' customs (Said, 1979). Having said this, the industry of expeditions and colonial voyages was ingrained in an imperial machine that imposed an essentialized version of the alterity as well as the importance of exporting the European model—at the best as the most successful model of mankind—to other cultures (Korstanje, 2018).

In a seminal book entitled *Colonialism-Postcolonialism*, Ania Loomba (2015) argues convincingly that we cannot understand correctly the colonial process without the counter-force, postcolonialism. Colonialism is neither a force introduced from the outside nor an economic project ended once the European empires left their colonies. It seems to be imposed and negotiated as something that occurs by the collusion of forces inside.

> 'Colonialism' is not just something that happens from outside a country or a people, not just something that operates with the collusion of forces inside, but a version of it can be duplicated from within. So that 'postcolonialism',

far from being a term that can be indiscriminately applied, appears to be riddled with contradictions and qualifications. (Loomba, 2015: 12)

Loomba triggers an interesting discussion around the paths opened by colonialism to the dialogue between two liminoid worlds. Colonial travels produced two types of marginalities: those who were stranded in their colonies with few possibilities to migrate to the urban metropolises in quest of a safer life and those who were enslaved and domesticated as a pariah in the Western societies. For the sake of clarity, she cites the example of African Americans whose ancestors were enslaved and exchanged as commodities in the US. As a global project, what remains clear is that colonialism effaced the precolonial world of natives, homogenizing unilaterally languages, ethnicities, and customs while imposing a one-sided version of history which can be scrutinized only through the lens of nationhood and the figure of the nation-state. The ***precolonial heritage*** seems to be drawn, almost always, by the ***history of colonialism*** never available to forms disengaged from the colonial reasoning. The knowledge production process, adjoined to travel writings, worked successfully and similarly to ideology, misrepresenting while restructuring a reality which locals finally internalized. Travels became possible to deepen a chasm differentiating between Europe and the rest of the world. In a nutshell, the non-Western Other concept resulted from the colonial discursive (Loomba, 2015). As Loomba (2015: 58) puts it,

> colonialism expanded the contact between Europeans and non-Europeans, generating a flood of images and ideas on an unprecedented scale. Europeans who travelled outwards took with them certain previous images of the people they expected to encounter. The actual encounters necessitated both the continuity and a reshaping of these images—continuity because previously held notions about the inferiority of non-Europeans provided a justification for European settlements, trading practices, religious missions and military activities—and reshaping in order to adjust images to specific colonial practices.

Doubtless, Loomba's contributions lie in the notion that traveler's stories and tales played a leading role amalgamating between fiction and entertainment and formed a culture that divided the inside from the outside. Such a division, of course, needed to form borders which were never static but were in constant negotiation. While the inside denoted a safe here, the outside implied an insecure there. In the same way, the self and the alterity were constructs in an ongoing shift. It is important to note that the history of colonialism is based on the logic of a representation of the "Other", a portrait which was fabricated and regulated by Western science. Novels and travel writings, in particular, not only reduced the "Other" to a spectacle, a type of show, but also stereotyped the non-Europeans as irrational (inferior) agents. The Western rationality, obviously epitomized in the light of science, marked the borders between superior and inferior agents. The needs of "being there", which travels articulated, confirmed a would-be dependency theoretically organized in the literature and arts. To put things bluntly, science and colonization were inextricably intertwined (Loomba, 2015; Pratt, 2007). The construction of the alterity in binary—opposed—terms is functional to its own demonization, or under-valorization. Ideology not only mediates between culture and reality but also says much for what it silences. In consonance with this, Nichole Guidotti-Hernández (2011) calls attention to the ideological disposition of language to legitimate an explicit form of violence over native. She coins the term *unspeakable violence* to denote the invention of history (and tourism consumption) to relegate some ethnic minorities to the periphery and transmit an imagined story about cruel facts. The nation-state disposed of—while subjugating—nomad tribes which were gathered and forced to live within the borders of the nation-state. As a powerful ideological concept, the sense of history only tells the story of the winners. This begs a more interesting question: is colonialism a necessary precondition for the rise and expanse of capitalism?

As Immanuel Wallerstein eloquently observed, the world-system model offers an interesting diagnosis to understand the material asymmetries of capitalism. In the same direction, the theory brilliantly punctuates the economic inequalities that originated in the colonial period perpetuated in the capitalist system. What is more important, the world-system theory holds that there is a much deeper transnational division of labor that cuts the world into core and semi-periphery and periphery countries. While the exemplary core develops on higher skills, labor-intensive production, and mobile technologies, specialized in travels (*transport technology*), the periphery countries play a passive role providing raw materials (commodities). Per Wallerstein's viewpoint, the origin of capitalism should be traced back to the fourteenth and fifteenth centuries—to be more exact between 1290 and 1450—when Europe took advantage of a technological revolution to conquer the Americas, laying the foundations toward the British industrial revolution. For Wallerstein, the origin of capitalism was certainly accidental. The unequal development, among nations, is explained by the priorities of European powers to gain rapid control over world economy. In this vein, science represents the valid source of knowledge that legitimates the dependency. This knowledge is structured according to a world system formed of small mini-systems. These world systems are seen as unities more complex than mere empires. Rather they are formed by a set of mechanisms which distributes the amassed surplus value from the periphery to the core. While the periphery provides the raw materials, the core returns the finished product. Having said this, the market acts as a necessary means for the core to exploit the periphery. The end of the system is marked by a temporal crisis which expresses the problem of accumulation of the capitalist system. As Wallerstein (1991, 1998, 2004) puts it, the problem of underconsumption is given by the fact that the downward harmonization of wages creates further surplus for capital-owners in a short term, but in the long term, this measure reduces the demand for the products. To wit, Wallerstein coins the term *knowledge movement* to reopen and question those stereotypes, discourses, and voices cemented in the old colonial process and now inherited by the capitalist hegemonic (linear) cosmology. The social sciences were based on linear reasoning that gradually shaped a specific relation between the West and its alterity. Knowledge was unilaterally produced by Europe and passively accepted by "non-Western Other". To be more precise, in Wallerstein's terms,

> The movement within the humanities came to be called cultural studies. The prevailing view within humanities had been that these were objective criteria of beauty, canons or aesthetic universals, which could be known and thought but not changed. In this sense, the world was homogenous. The new critics insisted that the cannons were simply self-serving and self-justifying criteria of a particular group. (Wallerstein, 2015: 3)

In consonance with Wallerstein, American economist Richard Heilbronner (1980) alerts that the human mobilities oiled the machinery of the technological revolution, acelerating the necesary social upward towards the configuration of the capitalist society. However, he finds that global capitalism would never be possible without Americas' conquest, which meant a foundational event that mobilized factories, commodities, and human souls in the quest of a new radical transformation. Travels boosted a set of stagnant economies which exploited slaves' bodies to amass a surplus. While technology innovated in material production, capitalism changed medieval days' institutional basis toward a mobile world. *New Yorker* media critic Douglas Rushkoff writes that capitalism evolves on the basis of the centralized authority cemented by the colonial rule. Modern corporations and businesses are supported not only by governments but also by trade laws. The ancient central (legal) authority emanated from the monarch remained firm in the spirit of colonialism, especially in the sanction of new emerging monopolies where the European powers administer local trade. Colonialism consisted of an economic practice which legally prohibited the commerce with others; at the same time colonial powers monopolized and centralized the amassed wealth. European economies imported commodities from their overseas territories while returned *processed products* to the colonies. As a result of this, European powers have grown and expanded their economies to other regions of the world, installing capitalism as a global project (Rushkoff, 2000).

In addition, capitalism fabricated and offered a specific consciousness based on travels and the discovery (exploitation) of the "Other", reportedly "the non-Western Other". This coincides with what Anthony Pagden named as *hominem viatores* (travelling men). In his book *Lords of All the World*, Pagden describes brilliantly the first steps of conquerors as well as the narrative produced in the intellectual circles to legitimate or object the *conquest of the Americas.* Per his viewpoint, narrative discourses are historically recycled and adapted to current times. The voices of the past never interrogate unless by our own conception of what happened in the past. This means that the imperial logic looks at the present with the borrowed eyes of the past. Although the bulls of Alexander VI and the treaty of Tordesillas (signed in 1494) gave some legitimacy to Spain and Portugal over the Americas, no less true was that British and French philosophers of the caliber of Thomas Hobbes, John Locke, or Jean Jacques Rousseau exerted a caustic critique on the authority of the Pope (in Rome) to endorse authorities to servants who were not Catholic. The dispute was automatically closed when the scholastic school of Salamanca received news about the hostility of natives to offer *hospitality* to Spanish settlers. Per Pagden, these stories served as an ideological narrative that put natives beyond the natural right of free transit—and the rights of *hominem viatores.* Since natives were unfamiliar with the universal (natural) right that makes all men equal, it was clear that they were sub-humans. As a result, natives were unjustly but legally dispossessed of all their lands. The Western notion of hospitality, as least as it was applied in the formation of the modern nation-state, legitimated ideologically the *conquest of the Americas.* This reflects two important aspects of politics. On the one hand, ethics and the law often dissociate. On the other, institutions' meanings are politically manipulated to preserve the status quo (Pagden, 1995). From its inception, the sacred law of hospitality remained the touchstone of Western civilization (Korstanje, 2017a, 2017b). The point will be carefully addressed in the next section.

Hospitality and the West: A Difficult Relationship

We have carefully reviewed the historical role of imperial expeditions and science in the non-Western Other configuration, *an idealized figure functional* to the colonial rule. Now, we turn our attention to the importance of what ancient Greeks dubbed as *the sacred laws of hospitality* to the organization of the Western lifestyle. The subtitle shows certainly the difficult relation between hospitality, which assumes the fact of being open to the "Other", and the West, a civilization that transcended its own borders to colonize the world (Korstanje, 2017a). While hospitality signals a gift exchange process that cements a relational dialogue between hosts and guests, colonization operates in the opposite dynamic. In fact, Western civilization had serious limitations in understanding the presence of the *non-Western "Other"*. Here two significant questions surface: what is the connection between hospitality and Europe? And to what extent does the colonial heritage remain operating in our modern consciousness?

To answer these two question it is necessary to turn our eyes back to the seminal book *Of Hospitality*, originally authored by Jacques Derrida—arguably in dialogue with Anne Dufourmantelle. Per Derrida's viewpoint, hospitality can be offered or not, but beyond the rite lies a question which needs to be unraveled. The host always asks the guest, who are you and what do you want from me? The philosophical dilemma of hospitality often associates with the role of the stranger who questions us, shattering our world. The stranger (Xenos) represents an "Other" who speaks a foreign language, which paradoxically makes no sense to us. He tries to say something we do not understand. The translation mediates between lay-citizens and the Xenos (alien). For example, in Plato's dialogues, the stranger is presented as an "Other" who asks about Others. Starting from the premise that the world is understood through images which emanate from the language, the stranger splits the reality into two parts. The identity, far from being constructed by the family, derives from the inception of "Others". As Derrida eloquently explains, strangers have the real image of us simply because we tend to repress our undesired feelings. Likewise, the foreigner is scrutinized by the state when the borders of language are crossed. Socrates seems to be trialed and judged even if he does not understand the judges' language. In the rite of hospitality, guests are politically forced to adopt a foreign language and they have to abide by the local law. No one offers hospitality to those who lack a name or a history. Derrida invites us to imagine hospitality within the constellation of the rights of the foreigner. He distinguishes the foreigner as a stranger domesticated by the hospitality while the stranger remains a pariah, an undesired "Other" or a *parasite* (paraphrasing his terms). What seems to be equally important is that hospitality takes two contrasting connotations in Derrida's argumentation. ***Absolute hospitality*** is given to strangers without asking anything in return. Instead ***conditioned hospitality*** is conferred on those who can pay for it. In this token, Derrida gives an important step to dissociate how the nation-state deals with foreign tourists and migrants. But not only this, there is an important insight on the relation between hotels and hospitals. Whereas the former refers to economic exchange spaces where guests pay for the services, in the latter guests have nothing to give. But what does happen when guests breach the law?

Although tourists and migrants have different treatments, the private and the public spheres are protected by the law alike, as Derrida notes. The police reserve the right to monitor or intervene in a phone line if there are suspicions of a crime because hospitality simply dissolves before the commitment of a crime. It is noteworthy that Derrida toys with the belief that hospitality endorses foreigners of certain rights and duties. They are morally obliged to abide by the local law and customs. If this does not occur, ***hostility*** regulates the host-guest relations, keeping the order and the transgressed borders back. Not surprisingly hospitality and hostility share the same etymological origin (***hospitium***) (Derrida & Dufourmantelle, 2000). Lastly, hospitality goes in two contradictory directions: one by the affirmation of the social order through the respect for the law which makes all citizens equal, but at the same time transgressing (if not shaking) the reign of dogmatism imposed by the host. Hence aliens are often objects of curiosity and mistrust. This means that we are seduced but at the same time questioned by the presence of the foreigner (Derrida & Dufourmantelle, 2000). Daniel Innerarity (2017) reminds that ethics is the fundamental source of experiences of hospitality. He equates the stranger to the role played by the risk which is never welcomed. Our current obsession to achieve a zero-risk society leads us gradually to a climate of moral decomposition. Having said this, those societies which neglect the "Other" are simply doomed to exhaustion. Based on the Kantian legacy, theorists should not dissociate hospitality from ethics.

> The Idea of hospitality recalls the meaninglessness of claiming to establish definitive symmetry between what belongs to us and what belongs to others and between doing and enduring. We could understand the ethics of hospitality as ethics of events; this is more of a theory of passion than action because it relies on the experience that human life is less a set of sovereign initiatives than a set of replies to the invitations the world gives us, often without our consent. (Innerarity, 2017: 3)

In a nutshell, hospitality often escapes to our own desires while situating the invitation as the gateway that clearly stamps when a guest is desired or undesired (Innerarity, 2017). In earlier works, Korstanje and Olsen (2011) and Korstanje and Tarlow (2012) examined carefully more than a dozen plots of horror movies produced between 2000 and 2010. They found that after 2001 horror movies were characterized by a process of humanization of the evilness, which meant a radical transformation of rogues. Although this point will be re-visited in the following chapters, it is no less true that villains, who in earlier decades related to the animal kingdom like ants, bees, jaws, or spiders, are now people who look like us. This remark coincides with what Luke Howie (2016) calls *objective violence*. One of the durable effects of the war on terror has been the introduction of a new culture of fear where the "Other" who looks like us is portrayed as a potential killer, a terrorist who hates us. This sentiment embodies a discourse of radical evilness which can be observed in the entertainment culture. In the recent horror film plots, young tourists who are unfamiliar with the visited landscape are seduced to drink and eat. Nobody knows the real intentions of these evildoers who kidnap their hosts while they are sleeping. For the Western imaginary, the sense of evilness is given to the lack of hospitality. This happens simply because even the evilness, like the vampire or the demon, needs our invitation to step in (Korstanje & Olsen, 2011; Korstanje & Tarlow, 2012).

As a backdrop, hospitality occupied a central position in the construction of Western social imaginary and the design of the modern nation-state. Originally Pagden has given some insight on the relation between politics and hospitality, but indeed, he is interested in the ideological forms of conquest and knowledge than in hospitality-related issues. It is tempting to say that the discovery of the Americas showed two unparalleled changes. On the one hand, the Spanish and Portuguese conquerors came across a "non-Western Other" who did not look like their Asian hosts. This opened the doors to an increase of myths, stories, and speculations revolving around new customs practiced by these primitive men. Some of these habits were certainly awful for European taste such as cannibalism, kidnapping of women, or the reluctance to give strangers hospitality. All the narratives aligned with a new literary genre oriented toward domesticating the native. On the other hand, the Western project which ideologically vied for the appropriation of vast and wild landscapes was enhanced by the European man. Insofar as travels and expeditions were more than a type of entertainment for writers, travelers were mandated to document everything they gaze. In fact, Europeans looked to ways for inspecting new landscapes, territories, and tribes which were dully mapped as a possession of the Crown. It is important not to lose sight of the fact that even if scholars have devoted considerable time and efforts to discuss the history of hospitality, its limitations to be politically incorporated into politics remain out of any discussion (Still, 2006; Friese, 2009; Servelli, 2006; Korstanje, 2017b). In a seminal work, Heidrun Friese (2009) outlines the political contours of hospitality as well as the doctrine that revolves around the rights of traveling. We are socialized in a society where hospitality is based on the religious charity, which means a type of generosity for strangers. With the rise of nation-states, these discretional duties were legally imposed as inherited basic rights to all citizens. In consequence, the social spaces of aliens, strangers, residents, and citizens were socially re-drawn according to the notion of social justice. The derived tensions and disputes about *the absolute hospitality*—à la Derrida—obscure more than they clarify. To some extent, the nature of hospitality, which inscribes into the logic of gift exchanges, corresponds with a temporary sharing of space. While hospitableness ends when a stranger becomes a citizen, it is also true that second-born migrant generations are regularly considered as *undesired guests*. Beyond the rhetoric of the nation-state that engages with human mobilities lies the need to classify the first- from the second-class citizens (Still, 2006; Darling, 2014). Thinking in the hospitableness means interrogating the refuge's role, as a liminoid space of security and negotiation where migrants move. Hospitality always speaks of the ways and protocols of how strangers are treated (Shirazi, 2018). Of course, its meanings vary depending on the subject and culture, but refuge can be defined as a constant process of change where the notions of belonging and alienation are at stake. In recent years, migrants looked to inhabiting—probably in a precarious legal state—the urban cities, constituting what Jullie Young dubbed as *the practice of the city as-a-refuge*. Nation-state centers its authority on the legal scaffolding legitimating its sovereignty which excludes the undesired "Other". The state reserves the right to accept or expel the stranger at its discretion. Contrarily, the city as a space of anonymity offers a good opportunity to thousands of migrants and asylum-seekers (Young, 2011). This contradiction has been widely discussed by Zygmunt Bauman in his earlier texts. In Bauman's mind, the modern city reflects the great asymmetries and contradictions of the capitalist system. As a stranger living in the UK, Bauman argues that a new dynamic of *liquid modernity* has systematically disposed of a climate of extreme competence (among workers) and consumerism (among consumers). Based on the urgency of melting the solids, as he explains, capitalism reproduced conditions of mega-urban cities, where we all became strangers to ourselves. To wit, the political nature of hospitality derives from the possibilities of state to administer human groups as classes, ethnicities, or ages. In the global city, there are doomed and salved people. While tourists are legally encouraged to travel worldwide, migrants and asylum-seekers are condemned to poverty and exploitation (Bauman, 2001, 2004, 2011, 2017). What is the relation between nation-state and hospitality?

Of course, this is a question Bauman does not approach, at least directly, but he gives some hints to solve the puzzle. In ancient times, hospitality was a socio-political institution designed to keep an inter-tribal alliance that activated a joint defense in an external attack. Based on a reciprocal logic, hospitality allowed guests interchange in peace time. Hospitality was practiced by Germans and Iberians, Greeks, and Celtic tribes since 5 BC. Naturally, the Roman Empire was opposed to the practice of hospitality because it tended to a decentralized form of power, impeded to negotiate with them. Once the empire collapsed, a feudalization process extended through Europe but the hospitality was more alive than ever. Travelers coming from different corners paid 10% of their properties to the Catholic Church to be assisted at a hostel. These establishments welcomed not only outsiders but also sick people in search of medical assistance. The terms *hospital* and *hotel* have the same etymological origins. Let's remind readers that low mobility was a characteristic of the medieval times, and travels were dangerous and a discouraging option. Once feudal kingdoms unite to struggle against their neighbors, the modern nation-state succesfully consolidates.

Hospitality plays a leading role in marking the essence of natural rights and the ideological core of nation-state. As Pagden eloquently described, the conquest of the Americas is ideologically legitimated by the lack of familiarity of some tribes in offering hospitality. The consolidation of the nation-state and the expansion of imperial powers were needed from the invention of inferior alterity to cement their authorities. In doing so, the political intervention was successfully orchestrated in two domains. In the lines of an internal domain, many aborigines and different ethnic groups were forced to live together within the limits of a nation-state sharing the same language and heritage. In an external line, the idea of alterity enhanced a sentiment of belonging for the national ethos. Having said this, the invention of "the non-Western Other" alimented the Western social imaginary over centuries, cultivating the needs of "being there" to understand "the non-Western Other". The imperial travels, as was discussed here, aligned with an imperial matrix that originated to colonize overseas territories while posing the Western reasoning (scientific knowledge) as a clear sign of superiority over the colonized indigenes; some emerging disciplines like anthropology and ethnography alerted the importance of protecting the indigenous heritage before its final destruction. This rhetoric alluded to an *uncanny paternalism*, which paved the way for the rise of the archetype of the *noble savage*. In this period, hospitality situated as the symbolic touchstone of the West. As stated, Western hospitality, probably from its inception, has serious limits and problems to understand the "Other" beyond Western rationality, which historically domesticates and eradicates the difference. In this cosmology, the "non-Western Other" is only welcomed when he is subject to the European lifestyle, venerating its so-called superiority over other cultural forms. Those voices or persons who confronted the nostalgic paternalism of the West—such as the nomad groups—have been systematically effaced while culturally pushed into the dust of oblivion. While colonialism started cultural destruction, the need for returning to a biased or idealized past (heritage) emerged. Today, the globalization process has enlarged the borders of Western civilization to other non-Western confines. In fact, national borders have been blurred, accelerating an identity crisis which has certainly epitomized in the problems of global terrorism and COVID-19. The "non-Western Other", who traditionally was an object of curiosity, becomes now an ***undesired guest***. This slippery matter will be discuss in more detail in the following chapters (Korstanje, 2017a, 2017b, 2018).

Conclusion

In this introductory chapter, we put Western hospitality as the main point of entry in a debate that is open even to date. The chapter describes the sociological relation of travels and culture as well as the attention of social science to expeditions and travels to describe far away cultures. From the outset, the worlds of travels, expeditions, and social science were inextricably intertwined. The notion of the West has been built according to countless expeditions and scientific travels. Anthropology and ethnography resulted from the colonial travels that were oriented toward documenting, gazing, and drawing a new (Europeanized) world's cosmographies. As a form of knowledge, travel occupied a central position in the configuration of European audiences. From Marco Polo to Caesar, all of whom are considered as pioneering ethnographers, settlers documented their discoveries and conquests, appropriating the alterity in their respective texts.

The West constructed the notion of self as a prerequisite of its expansion, which means that the term *self* embodies "the rational exercise of the Western superiority" over other cultures. This also meant that other(s), if they are to get the same mobility rights, are granted from the West. This continues even now in the form of the expectation of the Western powers that every other country should let them travel and work visa free while citizens of those countries will have to go through numerous approval processes. The Western mind does not bother about the right of the Other(s) to travel even in their intellectual realms; its support for repressive political regimes that thrive upon limited political and intellectual freedoms of their citizens is a testimony to this.

Beyond the curiosity for the non-Western Other, the Occident developed an uncanny paternalism. The non-Western Other was considered a child-like savage who deserved to be re-educated according to European reason. If pre-modern Europe disappeared in view of the advance of modern reasoning (industrialism), in the same way, the "Non-Western cultures" will wither away. Social scientists devoted time to collecting many artifacts, literary pieces, customs, and archaeological material before the final exhaustion of aboriginal cultures. This notion which occupied a central position in modern anthropology—far from disappearing—seems to be present in modern heritage tourism. This chapter reviewed how the "construction of a non-Western Other" constituted Western civilization's ideological core. When European economies ushered in the urgency to expand (to colonize) other economies, the figure of alterity and hospitality played a leading role. In between two worlds, Western visitors began to recover the lost paradise exporting while unilaterally imposing their civilizing values worldwide.

References

Altheide, D. L. (2006). Terrorism and the politics of fear. *Cultural Studies, Critical Methodologies, 6*(4), 415–439.

Altheide, D. L. (2007). The mass media and terrorism. *Discourse & Communication, 1*(3), 287–308.

Ashcroft, B., Griffiths, G., & Tiffin, H. (2003). *The empire writes back: Theory and practice in post-colonial literatures.* Routledge.

Bandarage, A. (2019). *Colonialism in Sri Lanka: The political economy of the Kandyan Highlands, 1833–1886* (Vol. 39). Walter de Gruyter GmbH & Co KG.

Bassin, M. (1987). Imperialism and the nation state in Friedrich Ratzel's political geography. *Progress in Human Geography, 11*(4), 473–495.

Bauman, Z. (2001). *The individualized society.* Polity Press.

Bauman, Z. (2004). *Wasted lives: Modernity and its outcasts.* Polity Press.

Bauman, Z. (2011). Migration and identities in the globalized world. *Philosophy & Social Criticism, 37*(4), 425–435.

Bauman, Z. (2017). Tourists and vagabonds: Or, living in postmodern times. In *Identity and social change* (pp. 19–32). Routledge.

Blauner, R. (1969). Internal colonialism and Ghetto revolt. *Social Problems, 16*(4), 393–408.

Calvo, T. (1996). *Iberoamérica de 1570 a 1910* (Vol. 252). (Iberoamerica from 1570 to 1910). Barcelona: Ediciones Península.

Clifford, J. (1989). Notes on travel and theory. *Inscriptions, 5*(5.29), 11.

Clifford, J. (1997). *Routes: Travel and translation in the late twentieth century.* Harvard University Press.

Cooper, F. (2005). *Colonialism in question: Theory, knowledge, history.* University of California Press.

Cowen, M., & Shenton, R. (1991). The origin and course of Fabian colonialism in Africa. *Journal of Historical Sociology, 4*(2), 143–174.

Darling, J. (2014). From hospitality to presence. *Peace Review, 26*(2), 162–169.

Derrida, J., & Dufourmantelle, A. (2000). *Of hospitality.* Stanford University Press.

Eisenstadt, S. N. (Ed.). (1967). *The decline of empires* (Vol. 154). Prentice-Hall.

Friese, H. (2009). The limits of hospitality. *Paragraph, 32*(1), 51–68.

Fryd, V. G. (1995). Rereading the Indian in Benjamin West's 'Death of General Wolfe'. *American Art, 9*(1), 73–85.

Gellner, E. (1998). *The state of the nation: Ernest Gellner and the theory of nationalism.* Cambridge University Press.

Gibbon, E. (1887). *The decline and fall of the Roman Empire, Vol. I.* Dent.

Guidotti-Hernández, N. M. (2011). *Unspeakable violence. Remapping us and Mexican National Imaginaries.* Duke University Press.

Harris, M. (2001). *The rise of anthropological theory: A history of theories of culture.* AltaMira Press.

Heilbronner, R. L. (1980). *The wordly philosophers.* Penguin.

Hill, R. (2014). Internet governance: The last gasp of colonialism, or imperialism by other means? In *The evolution of global internet governance* (pp. 79–94). Springer.

Hobsbawm, E. J. (1954). The general crisis of the European economy in the 17th century. *Past & Present, 5*, 33–53.

Hoffman, P. T. (2017). *Why did Europe conquer the world?* (Vol. 54). Princeton University Press.

Horvath, R. J. (1972). A definition of colonialism. *Current Anthropology, 13*(1), 45–57.

Howie, L. (2016). They were created by man … and they have a plan: Subjective and objective violence in Battlestar Galactica and the War on Terror. *International Journal of Žižek Studies, 5*(2), 1–15.

Innerarity, D. (2017). *Ethics of hospitality.* Taylor & Francis.

King, A. D. (2015). *Urbanism, colonialism, and the world-economy.* Routledge.

Korstanje, M. E. (2017a). *Terrorism, tourism and the end of hospitality in the West.* Palgrave Macmillan.

Korstanje, M. E. (2017b). The globalization of fear. In *Terrorism, tourism and the end of hospitality in the 'West'* (pp. 127–151). Palgrave Macmillan.

Korstanje, M. E. (2018). *The challenges of democracy in the war on terror: The liberal state before the advance of terrorism.* Routledge.

Korstanje, M. E. (2020). Passage from the tourist-gaze to the wicked-gaze: A case study on COVID19 with special reference to Argentina. In B. George & Q.-u.-D. Mahar (Eds.), *International case studies in the management of disasters: Natural–manmade calamities and pandemics.* Emerald Group Publishing.

Korstanje, M. E., & Olsen, D. H. (2011). The discourse of risk in horror movies post 9/11: Hospitality and hostility in perspective. *International Journal of Tourism Anthropology, 1*(3–4), 304–317.

Korstanje, M. E., & Tarlow, P. (2012). Being lost: Tourism, risk and vulnerability in the post-'9/11' entertainment industry. *Journal of Tourism and Cultural Change, 10*(1), 22–33.

Lehning, J. R. (2013). *European colonialism since 1700.* Cambridge University Press.

Loomba, A. (2015). *Colonialism/postcolonialism.* Routledge.

Mishra, V., & Hodge, B. (1991). What is post colonialism? *Textual practice, 5*(3), 399–414.
Nodelman, P. (1992). The other: Orientalism, colonialism, and children's literature. *Children's Literature Association Quarterly, 17*(1), 29–35.
Pagden, A. (1982). *The fall of the natural man: The American Indian and the origins of comparative ethnology* (Cambridge Iberian and Latin American Studies). Cambridge University Press.
Pagden, A. (1995). *Lords of all the world: Ideologies of empire in Spain, Britain and France c. 1500–c. 1800* (p. 63). Yale University Press.
Pollock, B. (2004). From nation state to world empire: Franz Rosenzweig's redemptive imperialism. *Jewish Studies Quarterly, 11*(4), 332–353.
Pratt, M. L. (2007). *Imperial eyes: Travel writing and transculturation.* Routledge.
Rushkoff, D. (2000). *Life Inc.* Random House.
Said, E. W. (1979). *Orientalism.* Vintage.
Servelli, M. F. (2006). Viajes y viajeros coloniales por las Américas [Travels and colonial travellers through the Americas]. *En: Revista de Crítica Literaria Latinoamericana, 32*(60), 335–339.
Shirazi, R. (2018). Between hosts and guests: Conditional hospitality and citizenship in an American suburban school. *Curriculum Inquiry, 48*(1), 95–114.
Smith, V. L. (Ed.). (2012). *Hosts and guests: The anthropology of tourism.* University of Pennsylvania Press.
Still, J. (2006). France and the paradigm of hospitality. *Third Text, 20*(6), 703–710.
Tilly, C. (2018). How empires end. In *After empire* (pp. 1–11). Routledge.
Toynbee, A. (1951). *A study of history.* Cambridge University Press.
Turchin, P. (2007). *War and peace and war: The rise and fall of empires.* Penguin.
Wallerstein, I. (1991). World system versus world-systems: A critique. *Critique of Anthropology, 11*(2), 189–194.
Wallerstein, I. (1998). The time of space and the space of time: The future of social science. *Political Geography, 17*(1), 71–82.
Wallerstein, I. (2004). *World-systems analysis.* Duke university Press.
Wallerstein, I. (2015). *The world is out of joint: World-historical interpretations of continuing polarizations.* Paradigm Publishers.
Young, J. E. (2011). 'A new politics of the city': Locating the limits of hospitality and practicing the city-as-refuge J. *ACME: An International Journal for Critical Geographies, 10*(3), 534–563.

Mishra, V., & Hodge, B. (1991). What is post(-)colonialism? *Textual Practice*, *5*(3), 399–414.
Nodelman, P. (1992). The other: Orientalism, colonialism, and children's literature. *Children's Literature Association Quarterly*, *17*(1), 29–35.
Pagden, A. (1982). *The fall of natural man: The American Indian and the origins of comparative ethnology*. Cambridge Iberian and Latin American Studies. Cambridge University Press.
Pagden, A. (1995). *Lords of all the world: Ideologies of empire in Spain, Britain and France c. 1500–c. 1800*. Yale University Press.
[illegible]
Pratt, M. L. (2007). *Imperial eyes: Travel writing and transculturation*. Routledge.
[illegible] (2000). [illegible] Random House.
Said, E. W. (1979). *Orientalism*. Vintage.
[illegible]

[illegible]

CHAPTER 2

The Colonial Voyages

Introduction

In the field of philosophy, Europe witnessed a radical struggle between voices like Jean Jacques Rousseau and John Locke. They condemned Western civilization's material benefits, valorizing the moral virtues of savages and those Hobbesian works that highlighted the urgency to establish a durable peace to avoid the *War of All Against All*. For Rousseau (2018), man is only good when he lives in the state of nature. Hobbes (1914) goes for the argument that the same form of nature only facilitates the conditions for a mutual eradication, lest by the imposition of a Leviathan reserving the use of force. When this occurs, the fear, which mediates among all men, mutates and sublimates into a new stage of civilization. Far from evaporating in the air, fear remains in the core of politics to keep the society united. The same social contract which ushers man into corruption in Rousseau ensures political stability in Hobbes (Bertram, 2004). Quite like travel, social contract occupies a significant position in the construction of Western imaginary as well as the literary circles (Morris, 2000). With an emphasis on an emerging literary genre known as *captive writings or novels*, aborigines become objects of cult and fear that align with the construction of modern European ethnocentrism. Captive writing was nothing new for the eighteenth and nineteenth centuries. Let us clarify to the readers that Miguel de Cervantes Saavedra has written numerous books and

M. E. Korstanje, B. George, *Mobility and Globalization in the Aftermath of COVID-19*,
https://doi.org/10.1007/978-3-030-78845-2_2

texts as a result of his experience as a captive by the Mors in Argel (1575). The foundational event emotionally marked Cervantes' life, who did not hesitate to denounce all cruelties and injustices he went through during his captivity (Garcés, 2005). Captive writing often enumerated a set of risky situations, acts of violence, or humiliations. European travelers were in the hands of hatred-filled aborigines (Klarer, 2019). In the constellations of this literary genre, Europeans were equated to a *tragic hero* who renounces everything in pursuit of an ideal, the protection of his nation (Strong, 2018; Biber, 2005).

The European host neither is a pure native nor comes back to life as a pure European citizen between two contrasting worlds. Metaphorically speaking, he is forced to live in a liminoid world (Benhayoun, 2006). Without any doubt, captive writing played a leading role as a precondition to travel writings, which illuminated and entertained the European readerships during the seventeenth, eighteenth, and nineteenth centuries. Although captivity writings take many forms depending on the century or the culture, without mentioning the differences between British and American captivities, they all refer to a forced journey, which means the urgency of being abducted from the family to live with strangers. Whereas the British sub-genre focuses on ethnographic potentialities, the American one centers on the spiritual logic of those who redeem themselves for love. Whatever the case may be, both connote the risks of traversing the borders of civilization to cope with savages in a foreign landscape (Snader, 1998; Voigt, 2009). The invention of a *radical Otherness* is vital to understand the process preceding the colonial expansion as well as the nature of captivity-based novels. These experiences have notable dissonance given the multicultural aspect of each travel while they were organized into an all-encompassing version where the European writers represented Indians, natives, or savages as dangerous aliens. However, this raises a philosophical dilemma: if aborigines were not strangers but natives of their lands, why do Europeans represent natives as strangers? And if so, strangers for whom?

These ideological tactics coincided with what Pagden (1995) named as semiotic strategies of domination and dispossession. The colonial expansion needed political and ideological legitimacy. In some cases, Europeans adopted a biased version of history. In contrast, in others, they re-founded the discovered lands, not only imposing their law and legal jurisprudence—without mentioning religion and customs—but making aborigines real strangers for the new order. Each founded city resonated with its

European counterparts; in this way current Mexico was baptized as New Spain by Spanish conquerors, while New York was originally named New Amsterdam by the first Dutch merchants, a term that was relabeled as New York by the Duke of York in England. New Orleans in Louisiana cannot be imagined without the city of Orleans in France. The city name relates to a much deeper mythological act of re-foundation, which is finely inherited from power exercise. When this happens, the locals become outsiders (Korstanje, 2018). Behind the act of discovery and dispossession lies the need to exert power over others. However, in doing so, a much-repressed sentiment of culprit surfaces. This sentiment is sublimated in different forms, which epitomize the invention of the radical "Other". We coin the term *ethnocentric hospitality*, which is pitted against *ancient hospitality* to connote the bestowing of rights to those aliens who align with European cultural values. At the same time, those who reject them are systematically eradicated.

The Invention of the Radical "Other"

For the Western social imaginary, non-Europeans were defined as a part inherently enrooted in the natural world, which is situated in opposition to culture. Like animals are part of a nature dissociated from the human order, aborigines keep a relational perspective respecting their environment. Instead, Western civilization breaks this relational logic with the animal kingdom, imposing a discourse based on eradicating a new re-building. The invention of the chair separates the modern man from nature, laying the foundations to a new radical cosmology to understand the "outsider". This distinguishes the modern man, who is designed by the Gods to administer the universe, from the hunter-gatherers. The binary nature of man is culturally enrooted in the genesis of Western (capitalist) thought. At this point, Tim Ingold introduces us to a more interesting argument. Western (capitalist) society is erected according to a binomial tension between them and us, work and leisure, here and there. Unlike hunter-gatherers who move freely in a geographically relational way, Europeans need the invention of a map to travel. Europe evolves according to a *dwelling perspective* which destroys the environment to build on. Hunter-gatherers go through the territory using their emotional memory, but Westerners are simply strangers who have no link to the visited soil. They need unique and novel experiences. The map serves as an instrument of guidance not to lose in the unknown land (Ingold, 2000).

In recent years, leisure travels have made a significant dent in the modern social imaginary as a sign of economic prosperity and civilization. The Occident successfully innovated in a technological revolution that led to exploring the world. Travels and colonial voyages were seen as the inevitable corollary of Western progress and supremacy over other ethnic groups who were historically—if not racially—relegated from European civilization's excellence (Blunt, 1994; Mancall, 2006). To put it differently, travels and expeditions were incipient capitalist Europe's bulwarks (Hadfield, 1998). What is equally important, as child-like (irrational) "Others", non-Europeans are often objectified as commodities to gaze. The rational and educated man is legally encouraged to travel as a colonial subject worldwide. His gaze is objective and speaks of the outside and the hostile world (Pratt, 2007; Loomba, 2007). In this respect, Ania Loomba (2007) argues that European empires devoted resources to consolidate their hegemonies over the natives by eradicating local cultures that they simply considered hostile to Western advance. To fill the gap, Europeans imported the idea of nation-state and nationhood, which have helped centralize resources, but paradoxically at the same time, they implanted in the natives' idiosyncrasy the dichotomies between us and them or civilization and barbarism. The binary posed by Europeans not only prompted the decolonization process just after the end of World War II but empowered local natives to shape their anti-colonial sentiments, which were organized to eradicate the European influence from their hearts (Loomba, 2007). This inevitable tension, which resulted in anti-imperialist and nationalist narratives, affirmed the architecture of the nation-state as a dominant project while cementing the tenets to the consolidation of global capitalism worldwide. The notion of a radical alterity seems to be vital to understand the colonial period's success preceding the capitalist system (Bear, 1998; Korstanje, 2012). Paradoxically, at the time the national borders are crisscrossed by the industry of leisure travels, connecting the world in a global ethos, more resentment and fear of the non-Western "Other" arise (Korstanje, 2016).

As previously argued, the radical alterity problem occupies a central position in the expansion of captive writings and the invention of a radical "Other" which escapes to the archetype of the noble savage. This radical alterity does not want to converse with the Western traveler he looks to seduce and kill under vulnerable conditions. As an emerging novel genre, captive writing gradually evolves and consolidates in what Julia Kristeva called *perverse hospitality*. While the guest knows perfectly hosts have nothing to pay for the banquet, he plans to slave or exploit them as a form of payment (Kristeva, 1991). As Roy H. Pearce (1947) notes, captive writings seem to be more than a mere literary genre that originated to entertain and educate the European readership. Based on Indian captivity, the genre illuminated historians' and politicians' paths in knowing further about *aborigines.* American Indians often seize captives for three main reasons: to use them as slaves, for ransom, or for exchanging for prisoners violently hosted by other tribes or even to sell (Van Der Beets, 1994). The Western imaginary paid consistent attention to captivity to mold an ideological narrative that had a double purpose. On the one hand, it justified the imperial practices of dispossession and exploitation, which were morally condemned by religion. On the other hand, it cultivated a set of different forms of entertainment which amalgamated into the figure of the European hero, who resigns his own life and comfort to expand civilization (Stout, 1983; Snader, 1998). In this token, Pauline T. Strong (2018) recognizes that Europeans culturally embodied themselves as agents who struggle to domesticate nature and the hostile "Other" who originally is portrayed as savage, bestial, or demonic-seductive.

In their mythical struggle to civilize natives, the Europeans narrate the sad stories of those captivated by the Native Americans. Writers portrayed these hapless heroes undergoing humiliation and torture without mentioning acts of intimidation and continued harassment. In the US, the new colonists were compelled to represent themselves as survivors of captivity—especially among Indians—who provided the first steps in the foundation of a great nation. As part of Anglo-American literature, captive writing rapidly expanded as a widely consumed form of cultural entertainment, including films, games, songs, drama, and television. However, as Strong contends, captivity was not a monopoly of Indians but was practiced by both sides: European and Indian traditions. However, in American literary history and colonial literature, there is a selection of pseudo-events and stories where captors were essentially Indians. The most important elements in this game were the *captive self* and *the captivating Other.* Most of the victims were certainly innocent children or women who Indians reeducated to meet all their desires. The dilemma lies in the fact that captives start a rite of passage to convert themselves into savages. This discourse operates as an ideological instrument that puts the vulnerability of civilized Europe to move through the wilderness, of course calling the use of force as the only valid option. Strong goes on to write:

> In Scholarship, as a hegemonic representation, captivity has often been decontextualized such that it is viewed as a distinctively Indian practice than as a complex historical phenomenon affected in significant ways by the European colonial discourses and practices. Recently, however, ethnohistorians have begun to demonstrate the extent to which North-Eastern Indian patterns of warfare, captivity, and diplomacy were transformed as they met the demographic, political, economic and cultural challenges imposed by the European colonial expansion. (Strong, 2018: 7)

The captivity, echoing Strong, should be equated to a hybridized contact zone characterized by the conflictive encounters of empires, natives, and cultures. Such a literary genre served to paint broad strokes that illuminated Indians' ways, a material eagerly consumed by anthropologists and ethnographers as a reliable source of information. Nonetheless, captivity efficiently legitimates a much deeper logic of domination where the natives are ultimately typified as a radicalized "Other", stereotypes mainly cemented by science. To cut a long story short, captivity epitomizes a transformative operation where the seized "Other", who remains in a liminoid zone, is re-socialized according to his or her captors' cultural values. This means that Europeans who originally resist the wilderness finally succumb to become savage people, living and looking like them. The *captivating savage* humiliates while seducing the *captive self*, which exhibits a relational aspect of identity that leads to making from the stronger "Other" (the captor) a type of master (Strong, 2018). It is essential to add that the ruling elite launches to control the classificatory system of documentation, encouraged by travels, to re-map a new civilization where European writers culturally paint the "Other" as a cannibal devourer or a natural enemy of Western civilization. The urgency to document the environment eloquently equates to the needs to control what the European order esteems as a potential danger (Van der Beets, 1994; Pratt, 2007). In this respect, in a seminal book entitled *Caught Between Worlds*, Joe Snader (1998) argues the genre births associated with the needs of exploiting the rest of the world. The modern era witnesses the obsession for describing the suffering and subjugation in a foreign land. However, he disagrees with the belief that captivity situates in sync with the formation of the American character (and of course earlier than the publication of Rowlandson's story), going beyond a pre-modern origin in the Middle Ages or its variations in the threshold of time. Snader toys with the belief that captive narratives engulf in an intimate (if not imagined) experience to be in a *foreign culture.*

> The libertarian individualism of the captives' struggle, furthermore, often seemed to produce concrete economic results. In many a captivity narrative, the isolated protagonist emerged as a man of science, an entrepreneur, or an agent of colonial expansion, a figure of self-consciously modern rationalism whose intellectual skills could turn the abject experiences of foreign captivity into positive account. On the other hand, the captive's heroic potential often seemed to depend on an ability to immerse himself within the alien culture that detained him. (Snader, 1998: 59)

Transitionally, this literary genre enthusiastically goes from factual to fictional landscapes, alluding to the experience of captivity as a significant threat to the imperial order. Based on a colonial experience when the frontiers are unilaterally altered, the European civilization was interrogated further by savage people who were disposed of their lands by Europeans. Not surprisingly, moral values like civility, character, and freedom were at stake in countless of these texts. The fact is that "While some novels imagined captivity as a crucible for the development of a spectacularly successful colonial heroism, others imagined captivity as a sign of the potential dangers, failures, and iniquities of the colonial project" (Snader, 1998, p. 6). As a mirror of the cruelty of Europeans in their colonies, their acts are revised at home. This suggests that more the exploitation—or oppression—in the colonies, more are the interests in captivity and captive writing as a literary new genre. Originally the genre of captivity was distinguished from classic colonial travels. However, a closer look reveals precisely the opposite. Colonial expeditions center on a pejorative viewpoint of the native which was fed back from captive narratives (Snader, 1998); to be more precise, captive writings allowed the construction and reproduction of a radical "Otherness" who ambiguously was the object of curiosity (for science) and fear (for the colonial authorities). As a predecessor of travel novels, captive writings include different clear-cut stages. The chronicle situates actors in a British or European settlement which is attacked by Indians kidnapping some innocent women or children while killing the soldiers or men; a mythical journey beyond the settlement begins when captors decide to return to their tribe. The captive will desperately try to escape so that he or she can come back to the settlement; in other cases, European captives are mysteriously released by the captors (Snader, 1998).

It is safe to say that unlike the colonial voyages, which moved because of scientific curiosity, or the tourist travels mainly motivated by leisure purposes, these journeys were unilaterally performed by the captors (without the captive's approval). Hence, captivity narrative lovers are in the presence of a new element that plays a crucial role in the genre: *a forced journey* (Burnham, 1993; Potter, 2003; Voigt, 2009). Lisa Voigt eloquently argues that this forced journey places the captive not only in a difficult position, probably between the wall and the deep blue sea, but also in a liminoid space where the Indians eradicate the identity. The captive lives and feels in two worlds (intermingled between its past and future). The captive is not a European woman anymore because she was socialized according to the Indian values, but she is not a pure native because of her European past. Such an in between stage opens the doors toward an identity crisis that will accompany her for the entire life (Voigt, 2009). Like a tragedy which is imposed on the hero, captivity and captive writings reshape and re-found the basis of Europeanness in a foreign land. What is equally important is that once released the European woman is not accepted because her heart has been corrupted by the savages (Haberly, 1976; Friedman et al., 2001). Captivity and captive narratives stem not only from the needs of understanding the settlers' harsh experiences abroad but also from the allegories of an imperial discourse tended to create a radical "Otherness", a point which remains in the Western imaginary even to date.

Finally, captivity signals a new anthropological figure strictly associated with European colonialism, which packages the vulnerabilities of European travelers in a virgin and hostile soil; of course, nothing is said about the European expansion cruelties over the natives. The term *virgin* here is used in contraposition to the civilized (enhanced) space. Europeans were ideologically convinced that virgin spaces should be possessed to be labored by the grace of the Lord.

It is not strange that the captive maiden refers to virgin European ladies. Furthermore, this surfacing literary genre is originally designed to give an ideological basis that successfully explains why natives are, at least culturally speaking, inferior in comparison with Westerners. Although racial supremacy as a dominant discourse that legitimated the colonial expansion in the sixteenth and seventeenth centuries was gradually disappearing, a new emergent discourse based on cultural supremacy mounted as a leading paradigm in the successive eighteenth and nineteenth centuries. Travel writing continues this tradition, paving the way for the rise of new European ethnocentrism that originated to downgrade and domesticate the alterity—or at best to offer hospitality only to docile savages. Here the question is, how does the West treat those "Others" whose domestication falls short?

Travels, Discoveries, and Leisure Time

As discussed in the introductory chapter, historians have agreed that there is a distinction between *imperialism* and colonialism. Such dissociation escapes to the figure of domination because there were empires in every epoch humans dwelled. As Ronald Horvath (1972) explains, domination seems to be a concept very hard to grasp in some conditions. Americans accuse Russians of being an empire while Latin Americans blame that the US protects its imperial interests in the region. Some speculate that imperialism is partly enrooted in human nature. However, while imperialism is understood as a form of inter-group domination where few settlers are delivered to the colonies in the representation of the imperial office, colonialism appears to be linked to the voluntarily mass migration of settlers to a certain colony. Both have in common the interplay between an exemplary center and a periphery. The center-periphery relations are mainly based on economic or political dependency. The second element in question is travels which connect the imperial geographies in peace or war times (Hall, 2017). Also, travels mold collective memories and maps while moving reciprocal energy among the colonies and the imperial cities (Scobey, 2002). In a trailblazing book, *The Lonely Crowd*, American sociologist David Riesman (1950) argues that human groups differ according to three clear-cut characters which cyclically mutate according to the economic-led production. *Tradition-directed character* is typical of groups that venerate tradition while establishing long-time rules. The law and the chief's authority are two important aspects of these societies. Examples of this character are ancient cities (like Greeks and Germans) and aboriginal groups in the Americas. The *inner-directed character* obeys organizations where subjects live according to their internal mandate, organizing their behavior according to the family and religion's rules. Protestants and Mormons in the eighteenth century are examples of this subtype, according to Riesman . Lastly, the *Other-directed character* embodies a set of practices and cosmologies where the "Other" occupies a central position—as a source of identity. In modern (industrial) America, people make day-to-day decisions to gain the approval of the "Other". The Other-directed person needs from the Others esteem, if not love, to be valorized.

The Other-directed societies are eager to consume news or stories that happen in faraway territories. In these societies, science, travels, and journalism connect people to the alterity's presence, creating the conditions toward a climate of global consumption and trade. At the time that industrial societies expanded, the Other-directed character came to stay. This means an open landscape to consuming (gazing) other cultures and landscapes and trading with other economies. The quintessential feature of this subtype is, of course, *the nature of empires.* This point leads Riesman to a difficult position he wants to avoid. If empires transcend time, why does Riesman hold that Other-directed character is associated with industrial modernity? He tries to address this quandary by saying overtly that the Other-directed character is not limited to modern society. Other imperial structures like Babylonians and Romans have successfully developed this subtype in the past. Echoing Toynbee, he toys with the idea that human structures mutate and change cyclically, not linearly, from the inner to the Other-directed character.

As a backdrop, geographers and historians have worked hard to describe the cosmologies of travels that cement the imperial social imaginary. Amy Turner Bushnell and Jack Greene (Turner Bushnell & Greene, 2002) observe that the imperial geographies are mainly formed by *an ecumene* which symbolizes the immediate zone (of encounter) where the original settlers take control, followed by *a sphere of influence* that mediates between settlers and the native people who are pushed to live in peripheral areas, which are economically less favorable. The *core region* seems to be urban metropolises—geographically located in the colony—which connects directly to the central office where the monarch resides. As Turner Bushnell and Greene (2002: 3) stated, "the periphery existed in a symbolic but often uneasy association with the center. As we move from the center of society … in which authority is exercised … to the hinterland or the periphery, over which authority is exercised … attachment to the central value system becomes attenuated".

Irrespective of the economic background, the periphery should always be understood as a *zone of marginality* where the "Other" is undesired by the imperial core. People who move because of the wish to know other cultures or because they are simply eager to trade with natives often cross the frontiers to get a zone of influence, which means an area of exchange (hybridity). These zones remain beyond the control of the empire and out of the law (Turner Bushnell & Greene, 2002). Among the imperial actors, it is given a much stronger negotiation process that confronts further the paradigm of power (Turner Bushnell, 2002). In this respect, John J. TePaske (2002) acknowledges that European empires in the Americas materialized their mastery according to a medieval spirit, successfully importing medieval institutions of power and coactions. These institutions, which are orchestrated initially by the imperial office to domesticate the natives, keep a proper synergy to actively protect the central authority's interests. He coins the term *integral to empire* to denote political dialogue (harmony) among the different actors in colonial Spanish America, a legacy which per his viewpoint remains live to date. This begs an important question, when do imperial travels aim for the discovery change to become a leisure practice?

Here two assumptions are made. On the one hand, ancient civilizations like Romans, Assyrians, and Babylonians commonly performed leisure travels in search of unique experiences, something very similar to the current leisure travels or tourism. Of course, these travels were marked by the risks and the technology of the epoch. To set a clear and illuminating example, Romans had the tradition of conferring a leave of three months (*feriae*) to stimulate travels as well as visits to relatives. In this way, the empire assured a fresh dialogue between Rome and its colonies.

To some extent, it cemented the power of Rome over the other provinces wholly. In peacetime, thousands of citizens visited the imperial provinces while thousands decided to target Rome as an exemplary destination. The Latin term *feriae* continues to live in current languages like German (*die ferine*) and Portuguese (*das feriaes*) to characterize *holidays* (Korstanje & George, 2021). The point was somehow overlooked by tourism historians over years, believing that humans did not come across something like tourism until the Industrial Revolution. Archaeologists, today, have gathered important material to prove leisure travels are attached, as maidens, to the imperial structures (Carcopino, 2003; Thomson, 2010; Gates, 2011).

On the other, there was very well a passage from elitist travels to mass tourism, which merits some attention. In this vein, Daniel Boorstin (1962) laments that the expansion of modern tourism engendered a naïve spirit of discovery, which replaced the colonial expeditions. Tourists are compared to superfluous strangers who can travel. Boorstin uses the term *hyper-reality* to describe how media constructed the sense of reality, imposing pseudo-events while eradicating spontaneous facts. In his viewpoint, the real world is setting the pace for the image's empire; needless to say, the same applies to travels. Modern tourism and leisure travels gradually eradicate the meeting with the "Other", molding experiences to what visitors pay to gaze. The "Other" is tailored and sold (cannibalized) to meet all guests' desires. Boorstin's legacy not only shed light on different scholars such as Jean Baudrillard (2001) and Paul Virilio (1998, 2001), it also resonated with full force in American anthropologist Dean MacCannell, an authoritative voice in the sociology of tourism. Per MacCannell, modern tourism and leisure travelers should be understood as modern institutions engendered by global capitalism. Like the totem acts as a source of authority in the tribal organizations, likely enhancing social ties and tradition, tourism generates the necessary energy for society to not disaggregate. To put this in a different way, tourism acts as a catalyst institution that revitalizes the day-to-day frustrations by articulating staged landscapes. The concept of authenticity is notably vital to understand MacCannell's insight. Lay-people who are alienated by the capitalist ideology look forward to consuming authentic places and experiences, which are the background of leisure travels, but paradoxically they reach the opposite: a fabricated (pseudo-staged reality). Hence, the tourism industry activates worldly machinery of consumption (the tourist sight) that cannibalizes locals, cultures, and landscapes (MacCannell, 1976, 1992, 1990, 2001, 2002). Are leisure travels a type of banality of colonial expeditions?

Toward a Theory of Leisure Travels

The earlier section contained a question that inspires us in this section—so to speak, the nature of leisure travels and their connection to imperial colonialism in the threshold of time. To put the same bluntly, how did the conquest of the Americas started a process of cultural eradication, which crystallized in a tourist attraction in the centuries to come—a type of dark heritage where the culprits for the violent exploitation are never assumed. European tourists visit today's marginalized zones to affirm their loyalties for the modern nation-state that headed the massacre; they do not feel any remorse for the European states' sins (Tzanelli, 2016; Korstanje, 2020). Europe would never expand to colonize outside territories without a culture based on mobilities and an autocratic spirit of domination. As hotly debated in the introductory chapter, the West has evolved—as a civilization—according to ethnocentric hospitality where the "non-Western Other" should be domesticated because of the Western mainstream cultural values (Pagden, 1995; Quijano, 2007; Mignolo, 2007; Korstanje,

2017). For a correct understanding of leisure travels in the eighteenth and nineteenth centuries, we need to step back to the theory of hospitality.

The term *hospitality* comes from *hospitium*, a traditional ancient institution designed to engender inter-tribal solidarity. Tribes, which celebrated hospitality, coordinated efforts to keep a joint defense in case of an external attack. In the same way, in peacetime, they exchanged citizens while organizing events and banquets (Chapman, 1992). Hospitality occupied a central aspect of life in Germans, Celtics, and Greeks, in a nutshell in the ancient world. Of course, hospitality represented a serious political problem for Imperial Rome which struggled to impose its own institution, in particular *patronatus*, a centralized political authority in charge of a person (the emperor or someone delegated by him) (Wells, 2001). Although *patronatus* [patronage] has been unilaterally imposed to gain further legitimacy over the barbarian tribes, the hospitality has never been diluted, but it came to be operated by other silent means (Nicols, 1980; Sánchez Moreno, 2001). Ancient citizens strongly believed that disasters and famines were punishments disposed of by the Gods when any tribe failed to give hospitality to strangers. For the ancient voices, hospitality was a sacred law whose violation caused disaster. Not surprisingly readers may come across countless myths as Faunus or even Helen of Troy that narrate how the social maladies and disasters take a hit when hospitality is not honored. For some historians, this happens because ancient hospitality had a *relational nature*. Gods pretended to be vagabonds in quest of hospitality to test the human soul (O'Gorman, 2007; Lashley, 2008; Lynch et al., 2011; Jipp, 2013). In the same way, if the people protected strangers Gods will help them in the afterlife. Hospitality was, for ancient voices, *a sacred law*. There are plenty of myths and stories that tell us that only the evil or evildoers attack their guests.

Once Rome collapsed, hospitality was mainly managed by the Catholic Church. Each feud reserved the right to offer, or not, hospitality. By this time, Europe adopted a version of hospitality associated with conquest, protection, and political control. The medieval travels were perilous, and often, travelers needed the assistance of the Catholic Church at hostels. Travelers paid for the service 10% of their carried merchandise. Hostels received travelers and sick people in need of medical assistance (Pagden, 1995; Korstanje, 2017, 2018). Most likely, influenced by Rome, Europe made hospitality the touchstone of the universal (natural) right because hospitableness was culturally a concept that makes us equal before the law. We have mentioned in the preceding chapter how the Americas' conquest was articulated in view of the lack of familiarity of some tribe to honor hospitality. Here we come across a moot point previously referred to in the above-noted question. Is Western hospitality part of the problem or the solution to explain with accuracy the Western ethnocentrism?

The colonial process re-draws the borders of the new world, taking advantage of the unfamiliarity of the natives about Western hospitality. This expansion not only creates a sentiment of the culprit, which is filled with the needs of new literature like captivity writings, but sets the pace for a new political order where the *non-Western Other* is seen as an *inferior agent* who needs to be re-educated (Schorb, 2020). Having said this, the modern nation-state stimulates officers and travelers to re-draw the common borders. In doing so, colonial officers differentiate the external enemy while in the internal sphere indoctrinate countless nomad (aborigine) groups. Finally, the figure of captivity, which attracted a vast European audience, exhibited that the culprit's sentiment originated in the cultural eradication of the "Otherness". In the plot of these fictional novels, the hostile savage often embodied the negative features and practices Europeans adopted in the colonial period (Snader, 1998). This ambiguity, as mentioned by Snader (1998), allowed a double tactic. On the one hand, Europeans who originally transgressed their natural borders invading new virgin territories were seen as hapless victims whose open sin was to be there where they were not welcomed. The sin is finally expiated by invoking the European law (which emanated from the archetype of the colonial officer) that puts things in order with the use of force. The native is heavily punished because of his unfamiliarity with the law of hospitality. On the other hand, captivity was over-valorized as a necessary sacrifice for the glory of the empire and the consequence of natives' barbarism (Korstanje, 2017). The Western social imaginary has conceived the "alterity" based on binary connotation: the noble and the hostile savage; the former denotes any man who lived outside the European corruption (civilization) in harmony with nature and the environment, while the latter characterizes the animal brutality that escapes the control of rationality as well as the respect for the law (Todorov, 1999; Said, 1985). What is more important, the ancient hospitality sets the pace for a new, more ethnocentric version where the alterity is culturally framed to meet the status quo's values. In terms of Todorov, the European order understands that *human sacrifice* is associated with tyranny, but the massacre is not (Todorov, 1999).

Following the above-mentioned argument, it is important to add that *ethnocentric hospitality* applicable only to the docile "Other" sublimated the culprit originated in the European expansion. Needless to say, this novel literary genre that narrates the stories of captives seems to be the center of new travel or leisure travel literature (Mansfield, 2016). Echoing Riesman, the European empires opened the doors to control other overseas lands, and in doing so, they swallowed the natives. The decolonization process invited thousands of working migrants coming from former colonies to keep integrated into the European manpower. Capitalism successfully consolidates its hegemony at the same time that it expands its inner frontiers to colonize the entire world. Leisure travels offer a safe circuit for round-trips while excluding the modern international tourist map's wild zones. In a seminal book entitled *Economies of Signs and Spaces*, Scott Lash and John Urry (1993) talk about global capitalism, which cannibalizes cultures and human landscapes, exchanging them as commodities. In the same way, safe paths or cities are pitted against wild zones (risky destinations) often tourists avoid. Modern capitalism divides the world into two: the civilized safe towns and the dangerous rest of the world. As Korstanje (2017) puts it, the modern nation-state paradoxically eradicates nomad groups' real mobility to provide the lay-citizens a *tourist mobility* (closed to the Other). Natives are prohibited from engaging in warfare with neighbors, leading them to a cultural eradication. At the same time, they are obliged to live within new geographical frontiers. Natives and rural farmers are gradually pushed to live in modern capitalist cities whereas the development theory fills the gap between colonies and empires. As Phillips McMichael (2016) brilliantly describes, the end of World War II disarticulated the old colonial order, temporarily setting the pace for a new capitalist society. In this new world, Harry Truman called attention to the fact that developed economies should economically assist underdeveloped ones. The world was automatically divided into developed and undeveloped nations. Though financial assistance was provided by the International Monetary Fund and World Bank, far from placating poverty it aggravated it. Poor solicitant countries experienced not only economic stagnation but an increase of their debts. Authorities of these financial corporations consulted anthropologists and ethnographers to conclude that the cultural incompatibilities of solicitant nations explained the failures of development theory to adopt the rational values of Western economies. Likewise, the development paradigm cements a new European authority over former non-Western colonies, which Europe historically relegated to a peripheral position (McMichael, 2016; Korstanje, 2018). The long (colonial) racial discrimination which epitomized the old European order is replaced by a surfacing economic asymmetry widely legitimated by the development theory. Meanwhile, in the capitalist cities, workers often spend money (wages) in the sector of tourism or leisure travels, returning their surpluses to the *capital-owners.* Tourist travels are not only regulated by the law but workers are actively encouraged to travel country-wide. The first paces of modern tourism appear to be based on a dynamic initiated by the allegories of *honimem viatores* (the traveling men), which is consolidated by ethnocentric hospitality. It is noteworthy that tourism is not a modern institution. This leads us to think that there were ancient forms of tourism as well as leisure travels (not using that name) practiced by ancient empires. What we are allowed to say overtly is that tourism, similar to the industry of leisure, travels (regardless of its forms) should be understood *as the maiden of empires.*

Per Jacques Derrida (2000) absolute and conditional hospitality marked the difference between a mere utopia of those who ask for hospitality without anything in return and the political nature of hospitality. Even if Derrida is eloquently clear in respecting such a concept, he falls into tautological fallacy. Relational logic was a typical characteristic of ancient hospitality, which is endorsed by a religious spirituality, but the same never applied for *the ethnocentric Western hospitality*, reportedly a new version of hospitality where "the Other" is accepted under the condition that he can be controlled, stripped out of all his negative features. While ancient hospitality stresses the relational nature of being human, Western hospitality highlights foreigner's identity. Metaphorically speaking, *the Western Eden* reserves the right to open or close the doors of hospitality probably motivated by scrutinizing the stranger's biography as well as his credentials. Those aliens who are not aligned with the hosting community's cultural values are simply rejected as pariahs or, in terms of Derrida, parasites. We use the term *ethnocentric* here in the direction that the "Other" is never accepted lest he voluntarily embraces the Western cultural imprint. When this does not happen the stranger becomes an *undesired guest.* The theme will be continued in Chaps. 5 and 6 when we discuss the intersection of terrorism and hospitality. Ultimately, there is a continuum between museums and zoos. Both exhibit anachronistically *a different Other to the sightseers.* Zoos embody living places populated by animals, whereas museums offer the spectacle revolving around the inert dead, which includes utensils, arts and crafts, yields, books, and spears that belonged to aborigines or ethnographers. The past is idealized as a spectacle because it is never rememorized as it occurred. In sum, both orchestrate exhibitions of the inferior "Other". In the same way, the West expanded its frontiers and effaced entire aboriginal cultures at the same time a spectacle around the aboriginality has been mounted. At the time Western modernity killed real aborigines, it created a fancy recreation of the aboriginal world to be consumed by tourists (Korstanje, 2012).

Conclusion

Chapters 1 and 2 have critically discussed the role of hospitality and the technological breakthroughs in mobilities in the expansion the West has witnessed since the seventeenth century. Chapter 1 concentrates efforts in illustrating the cultural values that forged the imperial (colonial) rule, which configured the essence of the "non-Western Other". The urgency of anthropology and ethnography to document the aborigines' past paved the way for modern science's emergence without mentioning European paternalism. Michael Herzfeld (2002) calls the attention to re-discuss the hegemony of the West through the paradigms of what he dubbed as crypto colonialism, which coincides with the implicit exclusion of some countries from scientific knowledge, taking them as an object of curiosity (e.g., as an object of ethnographic research) but impeding them the access to the dominant instrument for economic prosperity. In this way, the past is unilaterally recovered but subject to the eyes of the present. Contrariwise, this chapter complements the idea that Europeans expanded to colonize the world, cultivating their own conception of politics and economics. However, this would never be possible without the invention of the figure of a radicalized "Other". This seems to be the lines of the contribution of this chapter. We argue that *captive writings* played a leading role in the configuration of new radical alterity that sublimated Westerners' sentiment of the culprit to pushing their frontiers. Even in the more contemporary Wild West movies, this is evident: the Westerner may be captured by a native tribe, may marry a native woman, may even accept to live the native way superficially, but this interaction either *reforms* the natives or is limited to low-impact minor aberrations.

Captivity writing was a precedent literary genre that captivated the curiosity of a vast European readership and molded *ethnocentric hospitality* that accepts the "Other" only when he meets the European mainstream cultural values. In this way, the sentiment of culprit that originated in the conquest sets the pace for forming an ideological discourse tilted to point out the cruelty of aborigines. The captive represented the archetype of a hero who sacrifices himself for the communal protection of his neighbors. If the ancient hospitality offered a relational logic where the vulnerability of the guest is never violated, the ethnocentric (Western) hospitality cannibalizes (if not dominates) the guest to display him as a morbid (tourist) spectacle. Of course, the colonial process opened the doors for the formation of new liberal economic forms of exploitation that energized global capitalism. However, the mistrust (fear) for the non-Western "Other" prevailed amply as the capitalist society's touchstone. In the seventeenth and eighteenth centuries, the tourist (leisure) travels had been framed as the Anglo-American empire's maiden. To wit, the tourism industry was gradually connecting the world, reducing costs and times of travels as never before. The technological revolution, which ignited significant innovations for mankind, accelerated the European expansion while connecting long-distant cities in hours. While the liberal state actively promoted values associated with democracy, economic prosperity, and free trade, strangers' fear increased. This fear appears to be related to an ethnocentric version of hospitality and the binary logic created by the radicalization of the alterity in the colonial period.

Travel writings remind that regardless of the education provided to the savage, he is still a savage, who sometimes can be controlled but not always. The current fear reigning in the West about terrorism or the advance of COVID-19 links to a sentiment of anxiety which leads us to be in alert waiting for the next attack of an enemy who lives within our society and of course who looks like us. Like the captive who was tortured or seduced by savages to abandon his civilization, the terrorist is seen as an "Other" whose Westernization failed. This happens simply because a captive will never be a good European man even if he returned successfully to his homeland. The argument will be fleshed out in the next chapters.

References

Baudrillard, J. (2001). *The uncollected Baudrillard*. Sage.

Bear, L. C. (1998). *Traveling modernity: Capitalism, community and nation in the colonial governance of the Indian railways* (Doctoral dissertation). University of Michigan Press.

Benhayoun, J. E. (2006). *Narration, navigation, and colonialism: A critical account of seventeenth-and eighteenth-century English narratives of adventure and captivity* (No. 17). Peter Lang.

Bertram, C. (2004). *Routledge philosophy guidebook to Rousseau and the social contract.* Psychology Press.

Biber, K. (2005). Cannibals and colonialism. *Sydney Law Review, 27*, 623–629.

Blunt, A. (1994). *Travel, gender, and Imperialism: Mary Kingsley and West Africa.* Guilford Press.

Boorstin, D. (1962). From traveler to tourist: The lost art of travel. In *The image: Or what happened to the American dream* (Vol. 77, p. 117). Vintage.

Burnham, M. (1993). The journey between: Liminality and dialogism in Mary White Rowlandson's captivity narrative. *Early American Literature, 28*(1), 60–75.

Bushnell, A. T. (2002). Gates, patterns, and peripheries: The field of frontier Latin America. In C. Daniels & M. V. Kennedy (Eds.), *Negotiated empires: Centers and peripheries in the Americas, 1500–1820* (pp. 15–28). Routledge.

Carcopino, J. (2003). *Daily life in ancient Rome: The people and the city at the height of the empire.* Yale University Press.

Chapman, M. (1992). *The Celts: The construction of a myth.* Springer.

Friedman, Y., Edgington, S. B., & Lambert, S. (2001). Captivity and ransom: The experience of women. In S. Edgington & S. Lambert (Eds.), *Gendering the crusades* (pp. 121–139). Columbia University Press.

Garcés, M. A. (2005). *Cervantes in Algiers: a captive's tale.* Vanderbilt University Press.

Gates, C. (2011). *Ancient cities: The archaeology of urban life in the ancient near East and Egypt, Greece and Rome.* Taylor & Francis.

Haberly, D. T. (1976). Women and Indians: The last of the Mohicans and the captivity tradition. *American Quarterly, 28*(4), 431–444.

Hadfield, A. (1998). *Literature, travel, and colonial writing in the English Renaissance, 1545–1625.* Oxford University Press.

Hall, C. (2017). Going a-trolloping: Imperial man travels the empire. In *Gender and imperialism.* Manchester University Press.

Herzfeld, M. (2002). The absence presence: Discourses of crypto-colonialism. *The South Atlantic Quarterly, 101*(4), 899–926.

Hobbes, T. (1914). *Leviathan.* JM Dent.

Horvath, R. J. (1972). A definition of colonialism. *Current Anthropology, 13*(1), 45–57.

Ingold, T. (2000). *The perception of the environment: Essays on livelihood, dwelling and skill.* Psychology Press.

Jipp, J. W. (2013). *Divine visitations and hospitality to strangers in Luke-Acts: an interpretation of the Malta episode in Acts 28: 1-10.* Brill.

Klarer, M. (Ed.). (2019). *Mediterranean slavery and world literature: Captivity genres from Cervantes to Rousseau.* Abingdon.

Korstanje, M. E. (2012). Reconsidering cultural tourism: An anthropologist's perspective. *Journal of Heritage Tourism, 7*(2), 179–184.

Korstanje, M. E. (2016). *The rise of Thana-capitalism and tourism*. Taylor & Francis.
Korstanje, M. E. (2017). *Mobilities paradox: A critical analysis*. Edward Elgar.
Korstanje, M. E. (2018). *The challenges of democracy in the war on terror: The liberal state before the advance of terrorism*. Routledge.
Korstanje, M. E. (2020). The dark tourist: Consuming dark spaces in the periphery. In M. Korstanje (Ed.), *Tourism, terrorism and security*. Emerald Publishing Limited.
Korstanje, M., & George, B. (2021). *The nature and future of tourism: Post Covid 19 context*. Apple Academic Press.
Kristeva, J. (1991). *Extranjeros para nosotros mismos [Strangers for all us]*. Plaza & Janés Editores, SA.
Lash, S. M., & Urry, J. (1993). *Economies of signs and space* (Vol. 26). Sage.
Lashley, C. (2008). Studying hospitality: Insights from social sciences. *Scandinavian Journal of Hospitality and Tourism, 8*(1), 69–84.
Loomba, A. (2007). *Colonialism/postcolonialism*. Routledge.
Lynch, P., Molz, J. G., Mcintosh, A., Lugosi, P., & Lashley, C. (2011). Theorizing hospitality. *Hospitality & Society, 1*(1), 3–24.
MacCannell, D. (1976). *The tourist: A new theory of the leisure class*. University of California Press.
MacCannell, D. (1990). Cannibal tours. *SVA Review, 6*(2), 14–24.
MacCannell, D. (1992). *Empty meeting grounds: The tourist papers*. Psychology Press.
MacCannell, D. (2001). Tourist agency. *Tourist studies, 1*(1), 23–37.
MacCannell, D. (2002). The ego factor in tourism. *Journal of Consumer Research, 29*(1), 146–151.
Mancall, P. C. (Ed.). (2006). *Travel narratives from the age of discovery: An anthology*. Oxford University Press.
Mansfield, C. (2016). *Alternative modernities in French travel writing: Engaging urban space in London and New York, 1851–1986*. Anthem Press.
McMichael, P. (2016). *Development and social change: A global perspective*. Sage Publications.
Mignolo, W. D. (2007). Introduction: Coloniality of power and de-colonial thinking. *Cultural Studies, 21*(2–3), 155–167.
Morris, C. W. (Ed.). (2000). *The social contract theorists: Critical essays on Hobbes, Locke, and Rousseau*. Rowman & Littlefield Publishers.
Nicols, J. (1980). *Tabulae patronatus: A study of the agreement between Patron and Client-Community*. Walter de Gruyter.
O'Gorman, K. D. (2007). Dimensions of hospitality: Exploring ancient and classical origins. In *Hospitality: A social lens* (pp. 17–32). Advances in Tourism Research. Elsevier.

Pagden, A. (1995). *Lords of all the World: Ideologies of Empire in Spain, Britain and France c. 1500-c. 1800* (p. 11). New Haven, CT: Yale University Press.

Pearce, R. H. (1947). The significances of the captivity narrative. *American Literature, 19*(1), 1–20.

Potter, T. (2003). Writing indigenous femininity: Mary Rowlandson's narrative of captivity. *Eighteenth-Century Studies*, 153–167.

Pratt, M. L. (2007). *Imperial eyes: Travel writing and transculturation*. Routledge.

Quijano, A. (2007). Coloniality and modernity/rationality. *Cultural Studies, 21*(2–3), 168–178.

Riesman, D. (1950). The lonely crowd. In *A study of the changing American character.* Yale University Press.

Rousseau, J. J. (2018). *Rousseau: The social contract and other later political writings.* Cambridge University Press.

Said, E. W. (1985). Orientalism reconsidered. *Race & Class, 27*(2), 1–15.

Sánchez Moreno, E. (2001). Cross cultural links in ancient Iberia: Socio-economic anatomy of hospitality. *Oxford Journal of Archaeology, 20*(4), 391–414.

Schorb, J. (2020). Captivity recast: The captivity narrative in the long eighteenth century. In S. Velasco, T. Strought Gahl, L. Johnson, & M. Soto (Eds.), *A companion to American literature* (pp. 250–267). John Wiley & Sons.

Scobey, D. M. (2002). *Empire city: The making and meaning of the New York City landscape* (Vol. 79). Temple University Press.

Snader, R. J. (1998). *Caught between worlds: British captivity narratives in fact and fiction* (Vol. 1). University Press of Kentucky.

Stout, J. P. (1983). *The journey narrative in American literature: Patterns and departures.* Greenwood Press.

Strong, P. T. (2018). *Captive selves, captivating others: The politics and poetics of colonial American captivity narratives.* Routledge.

TePaske, J. J. (2002). Integral to empire: The vital peripheries of colonial Spanish America. In C. Daniels & M. V. Kennedy (Eds.), *Negotiated empires: Centres and peripheries in the Americas, 1500–1820* (pp. 29–41). Routledge.

Thomson, H. (2010). *Cochineal red: Travels through ancient Peru.* Hachette UK.

Todorov, T. (1999). *The conquest of America: The question of the other.* University of Oklahoma Press.

Turner Bushnell, A., & Greene, J. 2002. Peripheries, centers and the construction of early modern American empires. An introduction. In C. Daniels & M. V. Kennedy (Eds.), *Negotiated empires: Centres and peripheries in the Americas, 1500–1820,* 1–14. Routledge.

Tzanelli, R. (2016). *Thanatourism and cinematic representations of risk: Screening the end of tourism.* Routledge.

Van Der Beets, R. (Ed.). (1994). *Held captive by Indians: Selected narratives, 1642–1836.* University of Tennessee Press.

Virilio, P. (1998). Dromoscopy, or the ecstasy of enormities. *Wide Angle, 20*(3), 11–22.
Virilio, P. (2001). *Virilio live*. Sage.
Voigt, L. (2009). *Writing captivity in the early modern Atlantic: Circulations of knowledge and authority in the Iberian and English imperial worlds*. University of North Carolina Press Books.
Wells, P. S. (2001). *The barbarians speak: How the conquered peoples shaped Roman Europe*. Princeton University Press.

CHAPTER 3

Pay Now, Consume Later: The Power of Delayed Gratification

Introduction

Whether colonial travels directly contributed to the advancement of science is studied at length; relatively less known is the fact that modern tourism, both in its nature and structure, has its origins in the massification of colonial voyages. Argentinean geographer Carla Lois (2007) suggests that colonial expeditions were accompanied by many risks and imaginary monsters that placed European discoverers in jeopardy. These mythological monsters not only were mapped but terrorized the colonial expeditions, shaping a medieval social imaginary reluctant to travel abroad. The advance of industrialism, doubtless, made travels not only easier but offered a new cosmology revolving around new external worlds and landscapes to gaze.

With the advent of industrial capitalism, it became increasingly apparent for the capitalists that some mechanisms needed to be developed to recapture the working class's wages. Nothing else came closer to this as the design of the tourism system. Capitalism took the soul away from the working class, and tourism was presented to them as a way to regain it. This also ensured the return of some of the wealth to the capitalist class. It took centuries for pro-poor forms of tourism to develop. When this did happen finally, the capitalists soon aimed to internalize it by creating

M. E. Korstanje, B. George, *Mobility and Globalization in the Aftermath of COVID-19*,
https://doi.org/10.1007/978-3-030-78845-2_3

pseudo-equivalents that would make tourists superficially feel authenticity while offering them the comforts of an environmental bubble.

Like in the days of COVID-19, tourism emerged from the urgency of buying today while consuming tomorrow's logic. In consequence, tourism consumption re-drew the contemporary geography, merging borders, cultures, and people. As this chapter discusses, if the liberal market posed the ideals of paying now to consumer later, our psychological expectancies, fears, and hopes remain in a hypothetical future. The chapter reviews the main contributions of economists theoretically to define the liberal market as well as the nature and evolution of national liberalism. In such a task, it is important to debate Western rationality again, that is to say, the ethics of capitalist exploitation without mentioning the radical changes of economist theory that passed to valorize production to consumption. The future's value creates hypothetical settings to stimulate mass consumption, which is dealt with in the two last sections.

The Nature of the Liberal Market (State)?

Without any doubt, for the popular parlance, the term *liberalism* is strongly associated with the idea of the liberal state. Today, economists and sociologists over-valorize the benefits of liberal societies over other organizational forms (Oakeshott & O'Sullivan, 2014; Joppke, 2004). However, the history of liberalism and how it resonates in the architecture of the state remains in its infancy (Galston, 1991; Wallerstein, 2011). As William A. Galston notably observes, the philosophy of liberalism concentrates efforts on honoring some specific values such as diversity, prosperity, and pluralism, but to some extent, some of these values are contradictory when the scientist places them under the critical lens of scrutiny. Is economic prosperity based on the subject's individualism? On the one hand, the liberal state should be understood as *an arena* hospitable to the different ideologies, viewpoints, and human orientation while protecting individual rights as well as the right of free association. On the other, liberalism should be framed in the line of philosophies cultivated to fight the rise and expansion of authoritarian spirits. At this point, liberalism not only promotes democracy but also affirms the foundations of free trade. The liberal state is certainly based on liberal unity which is cemented by the history of Englishness. One of the fathers of American liberalism, John Locke, holds the opinion that the concept of property is protected by the individual aspirations, goals, or methods and the crafted labor each person invests to enhance the lands. Starting from the premise that the right of the property exceeds the interests of any government, the principle of ownership should not be arbitrarily changed (Locke, 1966). This conception was widely criticized by the young Marx who envisaged the principle of ownership as one of the legal instruments that facilitated workers' vulnerability. Through his theory of surplus, Marx showed evidently how good exchange and the liberal market amount to a set of legal institutions oriented to ransack the workers (Marx & Engels, 1989). Beyond this dispute, the history of political modern liberalism—which so to speak takes the original ideas of the liberal philosophy of the French revolution and the voices of some inspiring writers in the eighteenth century—should be traced back to the end of World War II (WWII) and the bloody failure of the fascist enterprise. The origins of modern liberalism or neoliberalism invariably come from the first encounter held by The Mont Pelerin Society (known as MPS). This organization formed by scholars, writers, and some influential businessmen devoted considerable time and efforts to stop any resurgence of fascism or authoritarianism in the West. The organization protected the freedom of speech, democracy, free market, and citizens' individual rights toward *an open society*.

History tells that nine leading voices received an invitation from Frederick Von Hayek to debate the future of liberal thinking because of the advance of obstructionist methods in the economy such as Marxism and Keynesianism. In 1947 the first meeting took place in a room at Hotel Du Parc in the village of Mont Pelerin, Switzerland. Of course, this society has greatly influenced modern economists of Milton Friedman's and James Buchanan's caliber. In this vein, the ideals of MPS include the political organization of society according to free trade and democracy, followed by the fundamental right to develop the private spheres as correct functioning of the market while setting the minimum standards to undermine the predatory forces of the economy (Hartwell, 1995; Mirowski & Plehwe, 2015; Cornelissen, 2017). As Thiago Mata and Steven Medema (2013) outline, the end of WWII marks a new epoch where economists become real celebrities, intellectuals whose opinions and diagnoses illustrate politicians' agenda. The public interventions of economists are enthusiastically revered as the word of God. The old values of humanism that characterized Colbertism or other economic circles in past centuries set the pace to a managerial or materialist viewpoint of the economy, emphasizing mass consumption.

> On matters of public importance, economists are hired by the state and by private interests, and all sides in dispute, to participate in communicative practices in which economists might have limited discretion, as public relation experts orchestrate campaigns and their materials. It would be too extreme to say that the neoliberal moment has erased autonomy from the voices of economists; rather its most striking feature is the increasing value attached to publicity and celebrity, as economists compete among themselves for contracts and attention. (Mata & Medema, 2013: 15)

There was no other epoch where economists deployed their rationale and methods to locate and eradicate part of the problems or limitations, which often impeded societies to develop a state of durable economic maturation. Economists appointed the figure of poverty as one of the principal evils the capitalist culture should extirpate (Dimand, 2013; Goodwin, 2013). The important rational models that allow the application of successful production instruments were granted by the role of economists as the new modern philosophers of the inter-war periods. Economists look at those impediments or glitches that originated in the economic cycle. Those factors labeled by experts as undermining risks to the economic order should be scientifically studied, which means using the methods of science (Godden, 2013).

In recent years, some voices have praised the benefits of democracy and free trade as the tenets of global capitalism bringing not only economic prosperity to nations but also encouraging the liberal mainstream cultural values to the world in a struggle with undemocratic cultures (Fukuyama, 2005; Huntington, 1996, 2000, 2004). It is important not to lose sight of the fact that capitalism proved to be a successful model to warrant political stability and prosperity worldwide, reducing the risks of despotism, populism, and autocracies (Popper, 2020), without mentioning the creation of extractive institutions which are fitted against the ideals of liberalism (Robinson & Acemoglu, 2012); but, of course, this represents a part of the real story. British historian Eric Hobsbawm (2020) widely evidenced that the credits and discredits of capitalism has fluctuated in the threshold of time. After the crisis of 1930 which whipped the Western economies, communism was seen by economists as a firm, all-catch, model, at the very least, until the Soviet Union's collapse. The capitalist system embraced as a desirable model for the economist only with the turn of the century. This means that the credibility of economic models changes according to their capacities to offer pragmatic answers to societies' economic obstacles.

Richard Hofstadter (1992) and Michael Burawoy (1982) elicited caustic critiques to the problems of accumulation of global capitalism as well as inter-class conflicts, although from very different angles. Hofstadter focused his attention on the role played by social Darwinism, which was originally resisted but later accepted by Protestant Calvinism. For the capitalist culture, the doctrine of survival of the fittest agents is undoubtedly adapted to survive the strongest ones. The archetype of the intrepid businessmen emulates ideals revolving around *the superman* of Nazi Germany. What is more critical, social Darwinism postulated that some human groups are situated as the evolutionary pyramid's rank. Anglo-Saxons proved to be at the top of evolution—above all other races—because of their advances in technology, production, and mobilities. Similar to how evolution marks the pace for the fittest organism, the free market would determine the survival of the strongest citizens. For Hofstadter, social Darwinism determines the competition of all against all while the ruling classes are the only winner in a game where the winner takes all. Rather, Burawoy (1982) concentrates efforts to study the inter-labor relations at an American factory. Based on micro-sociology, he ascertains how capitalist consciousness is formed by vertical production relations (not discourses). He fills a gap left by industrial sociologists since the 1930s—why do workers spend their time working harder. Methodologically speaking, Burawoy involves himself as a machine operator in a Chicago factory. His outcomes are at least illustrative; workers often consent to their exploitation through what he dubbed as *the illusion of choice*. Workers feel more motivated when they believe their work contributes to a greater goal. Capitalist labor management not only maximizes productivity but also engages workers into all-encompassing class consciousness. There is a dialogue or negotiation between workers and factory owners which merits discussion. Per Burawoy, workers unions—far from solving the dependency of workers on capital-owners—affirm the illusion of choice.

In an interesting book titled *Liberal Loyalty*, Anna Stilz argues that citizens' loyalty particularly gives the political unity of liberalism in the liberal institutions, not in the history of liberal thinking. To some extent, liberalism cannot be divorced from the project of the modern nation-state. Liberalism operates in the same nation in different ways, as she explains. On the one hand, liberal nationalists emphasize the political obligations to cement loyalty to the same political structure (membership). The nation is morally justified because people in the exercise of their rational faculties elect to take part in society. On the other, the same liberal voices alert that democratic institutions work on the basis of a sense of solidarity only given by the national culture. Thus, cultural nationhood is seen as the most successful project that cultivates the individual identity while molding freedom. Per Stilz, the main contradiction lies in the fact that liberalism tries to combine national reciprocity emanated from the coactions of the nation-state with the promotion of individual wishes. The problem aggravates with the rise of cosmopolitanism which envisages the export of liberal thinking to other cultures. As she writes:

> Traditional liberals are up against challenges from nationalists and cosmopolitans alike, and this poses a painful dilemma. On the one hand, they could concede … that the moral principles on which liberalism is based should apply globally to all individuals, without reference to their geographical location, and therefore that special obligations to our own states and compatriots are in fact morally unjustifiable. This would save the liberal claim to moral universalism but at the expense of adopting the cosmopolitan position. On the other hand, traditional liberals could admit that democratic institutions tacitly presuppose a cultural nation, which provides the real demarcating criterion for who may and who may not belong, and defines the group of persons to whom we have political obligations. This would save the particularity assumption, but at the expense of betraying liberal Universalist background principles. (Stilz, 2009: 19)

The contradiction mentioned above centers on the dichotomies of rights and duties, especially when they are not compatible with other cultures and nations. Lay-citizens are educated to respect the law and to be good citizens, sending their children to the school, paying taxes, or simply voting. But paradoxically, these citizens are not familiar with the human right violations committed by their states abroad. The juxtaposition of liberality with the modern nation-state's project comes here with some ontological limitations (Stilz, 2009). As the history of colonialism discussed in an earlier chapter of this book proved, citizens sacrifice—while abandoning their critical reasoning—their autonomy or real freedom while they devote their loyalties to the nation-state (Stilz, 2009). At a closer look, exegetes of cultural nationalism had many problems in explaining why democracy should be globally adopted by other cosmologies and cultures even when it contradicts the proper national values. This discourse opposes the founding cultural values of liberalism, respecting the right of people to organize them (sovereignty) following a local law and authority. Since traditional economists embrace democracy as an institution that protects individualism which cements common solidarity to avoid the war of all against all, liberal writers believe that freedom is the essence of nationhood. The stratagem of liberals corresponds with making global a particular institution that is proper of European idiosyncrasy (Englishness); thus, democracy and liberal values are exported and imposed as the best of feasible worlds (Stilz, 2009).

The problems of competition and accumulation have historically been enrooted in capitalism. As Paul Krugman (1996) says, classic economists focus their attention on international trade, suggesting to authorities the need to enhance competition with other economies. Far from being a sustainable model, international competition accelerates the decline of economies in the Global North and South. Krugman elicits considerable criticism on the *neoliberal agenda* which punctuates a whole portion of the economic problems of the US that solves increasing competition. He toys with the belief that the economic crisis is induced by a decrease in productivity as well as some decline in the line of production. Unionism and internal labor conditions play a leading role in the rise of wages, not competition, as this expert agrees. Far from solving the problem, competition aggravates it. The neoliberal agenda coincides with the liberalization of capital while imposing serious restrictions on human mobilities. We live in a world where capital flows worldwide while migrants face serious barriers to step in the US or Europe. Why does this happen?

Well, an answer to this question aims to the voice of classic economists who in the 1990s advised Clinton's administration to stimulate international competition in the US, probably without credible numbers or firm robust evidence. Contrariwise, Krugman analyzes the economic problems of the US with statistics and numbers. To the speculation that the rise of the third world entails the ruin of the US, Krugman adds that it is ultimately safe to hold that financial investment in the third world increases productivity, equating wages between developed and underdeveloped economies.

In practical terms, each nation imports and exports goods depending on the domestic economy as well as the adaptative capacity to produce what other countries buy. This selective productivity is conceptually opposed to competition theory. He concludes that the real threat corresponds with the inequalities generated by the so-called free trade in developing economies. While the US boasts that capital should freely flow to enhance developing economies, serious restrictions to workers' migration are cynically posed. Krugman goes on to write:

> Ideas matter. According to recent newspaper reports: the United States and France have agreed to place demands for international standards on wages and working conditions on the agenda of the next GATT negotiations. US officials will doubtless claim they have the interests of the Third World workers at heart. Developing countries are already warning, however, that such a standard is simply an effort to deny them access to world markets by preventing them from making use of the only competitive advantage they have: abundant labor. (Krugman, 1996: 67)

Krugman laments that his colleagues misunderstand the logic and nature of the marketplace as well as the architecture of the nation. Private firms need competition to survive, probably to enhance their profits while minimizing the cost of production, but this never happens with nations. Nations, no matter their productive capacity, are never going out of business. There are a lot of external and internal factors that equilibrate the problems of the economy. To put it differently, what some rich nations sell, other (not richer) nations buy and vice versa. International trade seems much complex than a simplistic competition with rivals; the fallacy of international competition has been culturally enrooted in the literature from time immemorial, but it emerged recently in the ink of some neoliberal thinkers, Krugman concludes. He coins the term *pop internationalism* to refer to the set of fallacies, stereotypes, and ill-founded believes oriented to say that the US is losing leadership in the world given the rise of late-comers or new rivals which compete directly with American products. As he notes:

> It is fairly easy to understand why pop internationalism has so much popular appeal. In effect, it portrays America as being like a corporation that used to have a lot of monopoly power, and could therefore earn comfortable profits despite sloppy business practices, but now is facing an onslaught from new competitors ... unfortunately, it is a grossly misleading image, because a national economy bears little resemblance to a corporation. (1996: 118–119)

This dominant paradigm claims that everything is different now that the US opened the economy to international investors. Krugman's text is fascinatingly clear about the negative effects of pop internationalism laying the foundations for a better understanding of some protectionist discourses in the US. Of course, Krugman compiled this book decades before the triumph of Donald Trump and his "make America great again!" or "America first!", but his reflections have validity and relevance in the present days.

Last but not least, Douglas Rushkoff (2009) critically punctuates that modern capitalism derives from the colonial exploitation which created colonies receptive to European factories. This experiment paved the way for constructing a new cosmology aimed at exploiting the "Other". Although the colonial order diluted, its autocracy and authoritarian spirit remained live in modern society. Still further, capitalism is changing our cosmologies, undermining social ties. Lay-people are prone to competing with the "Other" to achieve their goals—probably instead of cooperating. The architecture, roads, cars, and all fabricated manufacture are disposed to make a more solipsist world strictly based on mass consumption. In this respect, we are creating pseudo-environments carefully designed to nuance the daily frustrations. These environments which only work as forms of communication mediate between the self and reality. Lay-citizens do not engage in dialogue with the "Other" unless it is by the use of digital technologies. As a cultural project, capitalism radically shifted the modern man's psychology, leading him to a climate of fear and mistrust (theory of zero-risk society). As a result, people are not valorized for their skills, status, or education but only for their purchasing power. The modern citizen, following Rushkoff, is *individually wrapped* to consume everything at his pace. Democracy seems to be functional for the liberty to buy and consume. Here two questions surfaced: why and under what conditions have we passed from a market centered on production to one centered on consumption? And what role does the future play in the construction of modern marketing campaigns?

From a Society of Producers to a Society of Consumers

For centuries, economists were certainly reluctant to accept consumption as a virtue or good practice for sanitizing the economy. Neither medieval philosophers nor Colbertism valued consumption as a preferable behavior to be emulated by citizens (Mason, 2002). Citing Smith, Kenneth Boulding claims that consumption should be understood as the end of production (Boulding, 1945). In his seminal book, *The Theory of Leisure*

Class, Thorstein Veblen (1973) laments that modern economists overlook the importance of production over consumption in their works. Per his viewpoint, a new emerging class characterized by the consumption of leisure is changing the basis of modern capitalism. The leisure class molds a type of new conspicuous consumption based on simulation, snobbism, and status. To wit, Zygmunt Bauman's works give a broad stroke that illuminates the ways how mass consumerism has changed human agency. Bauman toys with Marx's statement ***all that is solid melts into air***. He coins the term ***liquid modernity*** to denote a new stage of capitalism where goods are fabricated to be easily exchanged and consumed no matter what their quintessential characteristics. In earlier decades manufactured products served lifelong while now they are disposables. The same happens with human relations (Bauman, 2000). Social ties have been undermined because of a liberal life, where the market and materialist values dominate. In this vein, consumerism has disposed of a hybridized identity where people are unilaterally commoditized as goods, exchanged and consumed by the market (Bauman, 2004, 2007). While the traditional capitalist economy yielded goods for amassing wealth, nowadays the capital has been liberated. In consequence, consumers have indefectibly become consumed objects. Having said this, Bauman argues convincingly that ***postmodernity*** associates with novel ways of production and consumption where the agency is molded to be consumed, but in so doing countless nets of knowledge and information are organized given the dissociation of the effects from the causes (adiaforization process). Through the spectacle of disasters, consumers feel special, given the possibilities of other mortals. This point creates a clear dissociation between modern rationality and emotions, which the marketing experts widely manipulate. The constant quest for novelty erodes mutual trust, weakening social reciprocity (Bauman, 2007). When this happens, fear emerges to mediate between lay-citizens and their institutions (Bauman, 2006). The recent technological advances have been applied to achieve a zero-risk society. The motives for privileged groups articulating countless mechanisms of control and surveillance are twofold. On the one hand, technology is used to view and control the "Other", especially if he is a danger to the community. On the other, it gives prestige and status to technology holders who can fortress their homes (Bauman & Lyon, 2013). Postmodernity, in Bauman's terms, goes into a climate of liquid consumption where some actors are actively engaged in traveling worldwide, while others are pressed off their vulnerability, subject to cruel immobility (Bauman, 2017).

Although his eloquence and clarity are particularly brilliant, for some reason Bauman does not explain the socio-economic conditions for the passage from a society of producers to that of consumers. He elides to make a historical sweep of modernity as well as the history of economic theory. It is important to note that his conceptual framework dates back to the globalization process. This happens because Bauman is strongly interested in interrogating further the effects—not the causes—of modernity. For the sake of clarity, the other part of the story is given by American historian Kathleen G. Donohue (2003). In her book *Freedom from Want*, she proffers a strong diagnosis to understand the obsession for a modern economy to stimulate consumption. She explores the dominant discourse of American liberalism to epitomize the ideals of a modern consumer. Donohue avoids a caustic criticism, like Hofstadter, while she embraces historiography as her main method. The review of specialized literature in the fields of the economy from 1880 to 1940 leads her to find that the society of consumers consolidated as a leading paradigm after the 1940s. For traditional economists, consumption was negatively considered as a pathology (or social malady) of the economy. Over the decades, production and savings were two valid factors of economic progress. So, what has changed in the academic circles from Colbert or Ricardo to date?

Classical economy overtly proclaims that the only way for a nation to increase wealth is at a second nation's expense. First and foremost, comparable benefits and advantages can be very well obtained by international trade, when the exports surpass the imports. For society to successfully prosper, production should drive the lines of the domestic economy. While production is seen as creating added value, consumption is understood, by the opposition, as the inevitable destruction of such added value. As Donohue explains, lay-people are prone to moral corruption when they fail to stifle all their desires for these traditional voices. By the same token, frugality and hard work are two of the vehicles toward economic development. By the end of the nineteenth century, the idea of consumption was introduced by those economists who emphasized the need to promote general well-being in society. The notion of consumption as a bad degenerative force of the economy was finally replaced by ***productive consumption***. For expanding wealth consumption, production was inevitably a new issue that scholars discussed in classrooms and conferences.

> Classical economist might chastise mercantilists for sacrificing the interest of the consumer to that of the producer, but the logic of their theories could

> lead them to demand similar sacrifice. Although classical economists made room for consumption, they nevertheless remained convinced that production drove that the economy and that consumption only followed production. But, if production drove the economy, then the best way to further general well-being was to increase capital—this is, to develop more machines and build more factories. (Donohue, 2003: 10)

In sharp contrast to the European social order, America gradually divorced from the traditional conception of the economy to embrace new original paradigms. This new country growing in expansion required large manpower (dotted by the European mass migration) as well as technological inventions to enhance its current means of production. But unless the existing asymmetries created by the means of production were to be reversed, theorists realized that further production entailed further poverty, at least for the traditional economists. An answer to the problem of poverty and well-being laid the foundation toward a new understanding of working-class ideologies. Starting from the idea that *cooperation* was the most efficient manner to improve well-being, American economists acknowledged that economy equated to a struggle among different stakeholders and agents, all of them moving ahead to impose their interests. An economy strictly based on producers ushered society into strong monopolies or what is worse the abusive government-led worker unions. To resolve this, consumers had to work together, forming legal associations to protect their interests. Consumers' cooperation became the tug of war of the modern American economy by the turn of the nineteenth century. For these emerging theorists, poverty was a social malady that needed to be eradicated. Consumption acts positively in two directions. On the one hand, it energizes the production process enhancing the means of production. On the other, it creates more skillful and virtuous workers who make daily decisions to consume the best potential goods. Educating citizens to consume, so to speak, coincides with constructing the groundwork for the expansion of a great nation.

> The skilled and intelligent workers, by contrast, had more elective wants. Because they could satisfy these wants with any number of different goods, high consumption workers were able to choose goods that nature supply abundantly and forgo goods that nature would supply only with difficulty. Thus, unlike the ignorant and inefficient workers, they were able to harness nature's productive capacities. (Ibid: 46)

The creation of added value and wealth depends much on the inherited knowledge as well as the synergy invested in the individual labor while removing the obstacles toward a thriving consumer-oriented system. One of these obstacles was the inter-class inequality and material asymmetries that led society to a stage of underdevelopment. With this in mind, economists gradually proposed a new paradigm where consumers—not producers—play a leading role in boosting local economies while patronizing the logic of businesses. As Donohue puts it, the Cold War and the 1950s witnessed how left-wing economists scrambled to impose consumption as the new cure of the material injustices and problems of American capitalism. Paradoxically, they believed in the society of consumers as an escape to the materialist exploitation exerted by capital-owners. The fear of poverty, which empowered the role of consumers to set back the obstacles of the production process, associated with the importance to construct a legal scaffolding to achieve a competitive market. Only the market, American left liberals allude, encourages a climate of necessary competence to avoid monopolies' tyranny. Donohue goes on to write with extreme clarity:

> The consumer's power, however, was not absolute. It depends upon two preconditions. The market has to be a competitive one, and the consumer had to be informed. Only in a competitive market would consumers be able to punish a producer who did not produce high-quality, low-cost goods by taking their business elsewhere. And only if consumers were informed about the products they purchased would they be able to distinguish high quality from low, good value from bad. (Ibid: 120)

By the late 1970s, the paradigms of the American economy have been fully radically changed. Economy transformed society, passing from a society of producers to that of consumers. The interplay of economic and humanist theories crystallizes in a pro-consumer trend that situates democracy as the legal precondition for well-being. When the liberal market gives liberty to consume the best goods at lower costs, democracy forges *a consumer identity* with the end of re-energizing durable political stability. By the end of the twentieth century, marketing and management devoted considerable effort, time, and money toward dissecting the consumers' mind (Donohue, 2003), which means their fears, hopes, experiences, and preferences. The emphasis on consumption—far from solving the existing inter-class asymmetries—aggravated the US's economic situation, especially after the crisis of 1930. The end of WWII forged a new stage of capitalism that paved the way for the rise of a global (expansive) order that ultimately connected the world. As Alan Warde suggests, democracy and capitalism evolved around the ideals of freedom for choice. This illusion seems to be greatly exaggerated by philosophers and economists. By prioritizing choice over other values capitalism gives the impression that any consumer should be seen essentially as a decision-maker who only prospers in democracy. In this way, consumption dilutes as a taken-for-granted reality (or universal value) all cultures and nations should adopt. A scientific investigation oriented toward discussing and defining consumption should not evade a critical insight on its nature without mentioning the socio-economic background the concept has evolved (Warde, 2017).

The Future Governs Us: Pay Now but Consume Later

As the genesis and socio-technical conditions of capitalism have been extensively discussed in the earlier section, we now turn our attention to the future figure as the hallmark of global capitalism. The founding parents of sociology envisaged a precise intersection of future and capitalist consumption. While Max Weber (2009) proponed his thesis of Protestant predestination to explain the future-oriented aspect of modern man, Sombart (1913) used the metaphor of the desert to illustrate the abstract nature of money. Erich Fromm (2001) postulated a psychoanalytical explanation about the role of predestination in modern capitalism. The Calvinist and Puritan traditions stem from authoritarian personalities that go to the future to domesticate their own thanaptopic drives. Like fascism and communism, capitalism develops an uncanny fear of freedom originated in the need to dominate the "Other".

As a backdrop, Jorge Fonseca (2015) describes how the state's main role is to ensure the stability of the capitalist system. The nation-state looks to repress the internal conflict to benefit the economic distribution and capital accumulation process. In moments of turmoil or crisis, states defend the interests of big corporations. Today, globalization allows an easy investment to get cheaper commodities, but if the conditions of production are somehow interrupted, nation-states intervene to grant the affluence of capital. The discourse of the future offers a fertile ground to cement the capitalist expansion and Western hegemony. French philosopher Jean Baudrillard (2006) coined the concept of the spectacle of disaster to denote the current obsession for a zero-risk society. He compares the capitalist world to the plot of *Minority Report*, a dystopian futurist film where Precogs, a set of new mutants who see the future, help the police to detain potential criminals even before the crime takes place. The capitalist culture fabricates pseudo-events in the future to impose policies that otherwise would be rejected. In other terms, the future governs the present through the invention of future events that will never occur in reality. In this vein, Jens Beckert uses the term *imagined future*, echoing Anderson's imagined communities originally. He explains that capitalism evolves according to fictional expectations which are the fundamental fuel that oxygenates the capitalist economies. Far from being conditioned by the past, the present is certainly molded by the capitalist culture's future. Decisions, risk perception, fears, hopes, or expectancies are all lived in a hypothetical future. The self-made decision not only can be predicted by employing sophisticated software but is suspended for future rewards.

> The capitalist economy institutionalizes specific pressures that enforce a temporal orientation toward future economic opportunities and risks. Only by closely examining these institutionalized pressures may we comprehensibly shed light on the role of actors' temporal orientation with regard to economic process. (Beckert, 2016)

Per Beckert's viewpoint, two combined mechanisms actively support these future-based orientations which are typical of capitalism: *credit* and *competition*. The extreme market competition pushes actors to maximize their gains while reducing their costs. This climate of competition interrogates the rational agent and forces him to move to the future to gain further opportunities. Complementarily, credit gives agents tangible resources toward future economic success.

Enterprises and lay-people are socialized to feel free to ask for loans, for engaging in innovative activities that otherwise would be never undertaken. This logic opens the doors to a process of structuration (creation) and de-structuration (destruction) that Schumpeter baptized as *creative destruction*. Agents move to compete with "unknown Others", looking for new profits and opportunities and imagining situations they are not living in daily life.

> Capitalist economies destabilize and stabilize themselves by continuously undermining their historical forms: firms relentlessly seek new profit opportunities, employees strive to build careers, and consumers hunt for new

> consumption experiences. Surviving and thriving in an environment in which current forms will not last demands that firm, employers and consumers be constantly oriented toward a future they cannot yet see. (Ibid: 5)

Besides, the *American dream* serves as the best example of the connection between the future and capitalism. The discourse of the future articulates temporal orientations, institutions, and dynamics disposed carefully to regulate the excess of uncertainty (*politics of expectations*) keeping durable performative effects in society. This coincides with Reinhardt Koselleck's theory of an open future. He observes that in ancient voices, the idea of the future was strictly closed to the apocalyptic theory, divinely designed if not kept in secrecy by God. The introduction of an open future that is always undergoing change is widely introduced by modernity (Koselleck, 1997, 2004). In consonance with this, Giddens (2013) calls attention to the notion of risk, which as a social construct finely enrooted in an open (abstract) future had no place in medieval times but is limited to modern capitalism. The necessity to predict hypothetical scenarios recreates expectations and anxieties in modern minds as never before. This occurs because capitalism and risk perception are inextricably intertwined. Modern rationality frames the decision-maker as a rational agent whose decisions open the doors to new unseen risks. These entire risks move in a hypothetical future and do not always take place as imagined. To this end, Fred Block (2018) offers an interesting diagnosis of the problem of the future in modern capitalism. *The capitalist illusion* believes that capitalism is a closed system that follows its logic and rules that may be actively obeyed not to sacrifice the current economic well-being. The exegetes of capitalism are of the opinion that the system successfully achieves stable institution because of workers' competition. For this illusion to work, the idea of the future plays an important role in naturalizing while affirming human exploitation. In a seminal book entitled *The Corrosion of Character* Richard Sennett (1998) asserts that the surface of new capitalism coincided historically with the dismantling of the welfare state, which protected counter workers' basic rights. This new global capitalism endorses responsibilities to citizens avoiding private corporations and the state for their future. Lay-citizens not only are forced to compete with others for a job, but they believe that their future depends on their abilities and skills to perform their fate. Dispossessed of all rights gained since the 1970s, citizens see themselves as co-managers of their success. When something goes wrong, the capital-owner is devoid of any duty. Modern workers not only discredit durable relations but also openly reject stable jobs which are seen as signs of lack of creativity and resiliency. Ideologically, capitalism cements the process of exploitation while introducing the illusion that success or failure strictly associates with their inherent workers' skills instead of the socio-economic conditions. This point begs the question, to what extent modern tourism appears in the history of modern Europe (especially, after the end of WWII) as a form of a new leisure practice directed to fabricate future experiences?

All-Inclusive Tours and Tourism

Over the years, historians have discussed and targeted Thomas Cook as the founding father of the modern tourism industry. Cook, originally, was not concerned with the creation of a new sector; he was a strict Baptist spiritually motivated toward taking people out of alcoholism. In doing so, he managed the first all-included trip and travel where patients were unable to carry any money. The all-inclusive tours, which took a central place in the incipient first steps of the industry, were innovative tactics to tackle the problem of alcoholism in modern industrial England (Cormack, 1998). For some inconvenient reasons that are not relevant here, historians of tourism systematically cover this dark face of modern travels' origins. Psychological frustration, not leisure, was the touchstone of modern tourism. *All-included tours* were the rule except for the first organized travels through England and Europe, even in undemocratic regimes (Walton, 2005; Semmens, 2005; Pack, 2006). Sasha Pack (2006) speaks to her readers on the history of tourism in Franco's regime, a process characterized by a rapid economic recovery after the civil war. Thomas Kaiserfeld (2010) describes the importance of all-included tours in Nazi Germany. These works directly contradict the hypothesis that the tourism industry needs democracy to survive. With the benefit of hindsight, the collected evidence suggests that tourism necessitates political stability to standardize wider forms of consumption to the whole society, laying the foundations for a high-mobility culture. Authoritarian regimes would develop a more ethnocentric spirit targeting local tourism than democratic societies, which are open to discovering other customs and landscapes (Pack, 2006). Whatever the case may be, the logic of paying now to consume later is an inherent part of the modern tourism industry's nature. To give readers further details on this, we need to discuss the philosophical nature of tourism, which is kept on its margins by the tourism epistemologists. Some hints are provided by sociologist Christopher Holloway who in 1981 published "The guided tour: a sociological approach" at the classic *Annals of Tourism Research*. He makes an innovative and exploratory study of tour guides, alternating direct observations with formal and informal interviews. Per his viewpoint, guides emulate a *dramaturgical style* that sometimes confronts tourists' cosmologies. Likewise, the guide should be understood as a culture-broker who patiently constructs a fictional relationship between hosts and guests. All that tour guides do are targeted toward satisfying their customers, increasing an involuntary state of dependency. One might speculate that when tourists travel abroad they feel further insecurities (like colonial captives), in which case the probabilities of bringing money decline. The opposite is equally important; all-inclusive tour packages would often be bought when tourists do not speak the language of the visited destination (Enoch, 1996; Chang, 2007). Following this assumption, Wong and Kwong (2004) offer a valid model (method) to find the key factors that guide tourists to choose *all-inclusive package tours* when they travel abroad. The model based on all-inclusive packages includes further security, guarantees of timely departures, and customer care services. Of course, the model has some problems associated with swindling when tourists do not receive the paid services or potential claims of *misleading advertising*. Hence, tour operators should communicate clearly giving consumers as much information as possible. Today, all-inclusive tours are the most popular traveling mode worldwide and one of the modern tourism industry's outstanding features. Based on a sample of 280 tourists who were permanent residents of Hong Kong, Wong and Kwong reveal the following axioms: all-inclusive tours are selected not only by the travel agency's reputation but also by mouth-to-mouth recommendations. Social trust plays a leading role in the process of decision-making. Secondly, individuals who travel with children or relatives prioritize all-inclusive packages over other sub-segments (i.e., solo-travelers). Thirdly, TV advertisement was considered a key factor by interviewees at the time of purchasing an all-inclusive package. Other studies suggest that tourists preferably look for direct flights instead of opportunities for visiting shops or safety concerns (Chen et al., 2019).

Although the literature is going in the direction of trust and safety, more than can be done regarding the principle "pay now consume later", which is principally characteristic of all-inclusive tours. As George Ritzer puts it, tourist consumption radically varies from other forms of service consumptions. Buyers pay for a service they will receive at a later day.

Ritzer coins the term *presumption* to denote the convergence between the production of a good and consumption. Particularly, tourism consumption is based on the needs of getting future experiences (Ritzer, 2015). To some extent, migrants and tourists share the commonality of reaching a new (unfamiliar) environment in quest of something new. Forced migration moves in quest of new opportunities and jobs while tourists travel for the pleasure of discovering new landscapes and unique experiences. A much deeper difference lies in the fact that the tourist itinerary seems to be carefully designed through a safe path while migrants are often subject to countless risks and dangerous situations (Korstanje, 2018). Both start a journey from their incompleteness. Philosopher George Santayana (1964) holds that, unlike animals which move for hunger or thirst, we—human beings—depart from what we lack. The self needs to travel to escape the humdrum routine or an adverse situation. But the act of traveling is, above all, an act of trust, which appeases the sentiment of discomfort that originated in the uncertainty. While touring, we are not only alone, we also enter unfamiliar territory. As Santayana notes, the colonial settlers found emptied lands, but this was not the luck of forced migrants who came across a newly hostile world. He distinguishes the figure of the discoverer whose stories are read by a vast audience and the vagabonds who stranded in their vulnerability are an object of mistrust and fear. Any travel starts with a very deep process of adaptation or rejection. Unlike discoverers or vagabonds, tourists are psychologically motivated to be in contact with unfamiliar stories, customs, cultures, and geographies. This interest is not cultivated by the abundance of knowledge for the external world but by the outsider's lack of consciousness. In contrast to Santayana, for Marc Augé, tourist travel is a type of displacement that never takes place in reality, a voyage that ends when it begins. At airports, tourists are stranded without any type of identity or tradition. They only rush to show their credentials at border check-up points or for earlier embarking. Augé dubs these anonymity spaces as non-places because the past is eradicated by consumerism and the excess of future. Hospitality, an ancient institution that marked the distance between here and there, or us and them, has been blurred within a new hyper-reality where the "Otherness" has been radically transformed. To wit, the classic dogma of ethnography and anthropology should at least be revisited. There is nothing like a "non-Western Other" to study; now "the Other" becomes our neighbor. After all, we are strangers to ourselves!

In Augé's development, the high-mobility culture explains the combination of an excess of future, technological innovations, and the eradication of history, which influenced new geography of human ties. Not only tourism originates in the bourgeoise society but tourism consumption needs stem from the social status to survive. This means that tourists need to differentiate from others. They need to feel outstanding experiences.. Originally in the first days of modernity leisure travels resulted from the need to reward hard work, so to speak, as a type of social recognition. This logic expands the tourism industry while commoditizing new territories and cultures. Tour guides gradually foment mobility, engendering closed stereotypes that neglect real human diversity. In fact, as Augé stresses, tourists are compelled to consume (gaze) an imagined (essentialized) "Other". The cultural maelstrom, described by the Freudian texts, expresses through the needs of traveling or for the fictionalization the tourism industry offers. If the colonial discoverer felt fear of the unknown stranger trying to expand his own European world to overseas colonies, the modern tourist extrapolates a closed view of the "Other" which is reproduced in society.

Modern tourists—far from initiating new trips—are doomed to travel toward fancy destinations to put this simply. He adds that the impossible journey is that travel we will not make any longer because the genuine sense of discovery has succumbed to tourist consumption. We are not gazing at new landscapes; we are now gazing at flat screens. The fictionalized destinations inscribe the real world's essence where the "non-Western Other" is simply neglected or eradicated. This is why migrants are considered potential terrorists, enemies of a closed order (Augé, 1995, 1997, 1999). The terror liberated after the attacks of 2001 has not resulted as a direct consequence of Al-Qaeda's terrorism but the expansion of Western borders (Augé, 2002). Augé's contributions help us think about not only the radical transformation of travels and human geographies in the threshold of time but the crisis hospitality is suffering today. Western civilization's outer borders have been expanded to the confines of the planet; thus, some social categories such as identity, heritage, custom, and culture that inspired the anthropological theory in the past are now commoditized as tourist attractions. This expansion blurred the contours of the past and the present times, hybridizing them into an extemporal society, where the "non-Western Other" is now our neighbor. In a nutshell, a stranger's fear sets the pace for a new sentiment of panic that emerged from our manifest impossibilities to easily detect the internal enemy.

Without any doubt, this seems to be one of the philosophical quandaries behind the obsession for modern terrorism, a slippery matter we approach in the following chapters.

Conclusion

Colonial travels have illuminated the rise and expansion of the diverse literary genre, among them captive and leisure travel novels. It is safe to say literature, knowledge, and travels are inevitably entwined. As discussed in earlier chapters, colonial voyages gave the foreground for the colonial travel, which created a radicalized image of the *non-Western Other*. In this chapter, we stressed the birth of modern liberal tourism, which originally centered on the bubble model. Tourism acted as a psychological instrument of revitalization while laying the foundations for a liberal market where the future replaced the present time. Prepaid tours or all-inclusive tours emerged as a good reason to protect and help recover people from alcoholism. When Thomas Cook organized the first tour, he was concerned with ensuring travelers did not have money in their pockets—to avoid getting drunk. This means that the concept of buying now to consume later was finely enrooted in the ideological core of modern tourism.

Like in the days of COVID-19, tourism emerged from the necessity of buying today while consuming tomorrow's logic. In consequence, tourism consumption re-drew contemporary geography, merging borders, cultures, and people. We place contributions of senior historians of capitalism and cultural theorists such as Donohue, Bauman, Urry, and Koselleck under the critical lens of scrutiny. The chapter introduced readers to the theory and evolution of capitalism before beginning to inspect the ways terrorism changed us after 9/11. To some extent, terrorism and COVID-19 have some commonalities which deserve to be investigated (Chap. 4). Last but not least, if the colonial expeditions mapped the world according to an imperial matrix recreating the conditions for natives' exploitation, it is no less true that capitalism—far from leveling the injustices of colonial oppression—affirmed the negative connotation revolving around *alterity*, while deepening an ethnocentric version of hospitality where "the Other" is a silent enemy looming to attack anytime and anywhere. Probably this is the intersection between terrorism and COVID-19. The former is a silent enemy who looks like us—probably a neighbor or a friend waiting for the moment for the next attack. COVID-19 embodies the same archetype, making all us potential terrorists, salient carriers of a lethal virus that jeopardizes the social order. The concept of the future is vital for understanding not only terrorism but the fears beyond the pandemic of the coronavirus disease today whipping out the planet.

The fear of the unknown does not mean people are running away from the unknown. Rather, it often increases the curiosity to know and master the unknown; it makes the unknown attractive and worthy of pursuit. In such scenarios, mitigation strategies come to place. Modern tourism offers a lot of inherently risky products. Some of these seduce tourists to places of sorrow and gloom; dark tourism is nothing but a formalization of this (Mitchell et al., 2020). During COVID-19, tourism strategies speak to safety but also invoke the deeply held desire in man to *walk over the fire without being burned*.

References

Augé, M. (1995). *Non-places: Introduction to an anthropology of supermodernity*. Verso.

Augé, M. (1997). *L'impossible voyage: le tourisme et ses images* [The impossible journey: Tourism and its images]. Payot & Rivages.

Augé, M. (1999). *An anthropology for contemporaneous worlds*. Stanford University Press.

Augé, M. (2002). *Diario de guerra: el mundo después del 11 de Septiembre* [War fieldnotes: The world after 11 September]. Gedisa.

Baudrillard, J. (2006). Virtuality and events: The hell of power. *Baudrillard Studies, 3*(2), 1–15.

Bauman, Z. (2000). *Liquid modernity*. Polity Press.

Bauman, Z. (2004). *Work, consumerism and the new poor*. McGraw-Hill Education (UK).

Bauman, Z. (2006). *Liquid fear*. Polity Press.

Bauman, Z. (2007). *Consuming life*. Polity Press.

Bauman, Z. (2017). Tourists and vagabonds: Or, living in postmodern times. In J. Davis (Ed.), *Identity and social change* (pp. 19–32). Routledge.

Bauman, Z., & Lyon, D. (2013). *Liquid surveillance: A conversation*. Polity Press.

Beckert, J. (2016). *Imagined futures: Fictional expectations and capitalist dynamics*. Harvard University Press.

Block, F. L. (2018). *Capitalism: The future of an illusion*. University of California Press.

Boulding, K. E. (1945). The consumption concept in economic theory. *The American Economic Review, 35*(2), 1–14.

Burawoy, M. (1982). *Manufacturing consent: Changes in the labor process under monopoly capitalism*. University of Chicago Press.

Chang, J. C. (2007). Travel motivations of package tour travelers. *Turizam: međunarodni znanstveno-stručni časopis, 55*(2), 157–176.
Chen, N., Masiero, L., & Hsu, C. H. (2019). Chinese outbound tourist preferences for all-inclusive group package tours: A latent class choice model. *Journal of Travel Research, 58*(6), 916–931.
Cormack, B. (1998). *A history of holidays, 1812–1990* (Vol. 4). Routledge/Thoemmes.
Cornelissen, L. (2017). 'How can the people be restricted?': The Mont Pèlerin Society and the problem of democracy, 1947–1998. *History of European Ideas, 43*(5), 507–524.
Dimand, R. W. (2013). 'Perhaps I'm a Don Quixote but I'm trying to be a Paul Revere': Irving Fisher as a public intellectual. In T. Mata & S. Medema (Eds.), *History of political economy* (pp. 20–37). Duke University Press.
Donohue, K. (2003). *Freedom from want: American liberalism and the idea of consumer.* Johns Hopkins University Press.
Enoch, Y. (1996). Contents of tour packages: A cross-cultural comparison. *Annals of Tourism Research, 23*(3), 599–616.
Fonseca, J. (2015). Large enterprises and corporate power. In *The world is out of joint: World-historical interpretations of continuing polarizations* (pp. 69–83). Paradigm Publishers.
Fromm, E. (2001). *The fear of freedom.* Psychology Press.
Fukuyama, F. (2005). Building democracy after conflict: 'Stateness' first. *Journal of Democracy, 16*(1), 84–88.
Galston, W. A. (1991). *Liberal purposes: Goods, virtues, and diversity in the liberal state.* Cambridge University Press.
Giddens, A. (2013). *The consequences of modernity.* John Wiley & Sons.
Godden, C. (2013). Observers, commentators, and persuaders: British interwar economists as public intellectuals. In T. Mata & S. Medema (Eds.), *History of political economy* (pp. 38–67). Duke University Press.
Goodwin, C. (2013). Walter Lippmann: The making of a public economist. In T. Mata & S. Medema (Eds.), *History of political economy* (pp. 92–113). Duke University Press.
Hartwell, R. M. (1995). *A history of the Mont Pelerin Society.* Liberty Fund Inc.
Hobsbawm, E. (2020). *The age of extremes: 1914–1991.* Hachette UK.
Hofstadter, R. (1992). *Social Darwinism in American thought* (Vol. 16). Beacon Press.
Holloway, J. C. (1981). The guided tour a sociological approach. *Annals of Tourism Research, 8*(3), 377–402.
Huntington, S. P. (1996). Democracy for the long haul. *Journal of Democracy, 7*(2), 3–13.
Huntington, S. P. (2000). The clash of civilizations? In *Culture and politics* (pp. 99–118). Palgrave Macmillan.

Huntington, S. P. (2004). Dead souls: The denationalization of the American elite. *The National Interest, 75*, 5–18.

Joppke, C. (2004). The retreat of multiculturalism in the liberal state: Theory and policy 1. *The British Journal of Sociology, 55*(2), 237–257.

Kaiserfeld, T. (2010). From sightseeing to sunbathing: Changing traditions in Swedish package tours; From edification by bus to relaxation by airplane in the 1950s and 1960s. *Journal of Tourism History, 2*(3), 149–163.

Korstanje, M. E. (2018). *The mobilities paradox: A critical analysis*. Edward Elgar.

Koselleck, R. (1997). The temporalisation of concepts. *Redescriptions: Political Thought, Conceptual History and Feminist Theory, 1*(1), 16–24.

Koselleck, R. (2004). *Futures past: On the semantics of historical time*. Columbia University Press.

Krugman, P. (1996). *Pop internationalism*. MIT Press.

Locke, J. (1966). Of property. *Privatization: Critical Perspectives on the World Economy, 1*, 61–71.

Lois, C. (2007). Mare Occidentale: el territorio atlantico en los mapamundis del siglo XVI—[Western Seas: Atlantic territories in the maps of XVIth century]. In P. Zusman, C. Lois, & H. Castro (Eds.), *Viajes y Geografias* [Travels and geographies] (pp. 33–50). Prometeo.

Marx, K., & Engels, F. (1989). *Collected works* (Vol. 31). International Publishing Company Incorporated.

Mason, R. (2002). Conspicuous consumption in economic theory and thought. In *Intersubjectivity in economics: Agents and structures* (pp. 85–104). Routledge.

Mata, T., & Medema, S. (2013). Cultures of expertise and the public interventions of economists. In T. Mata & S. Medema (Eds.), *The economist as public intellectuals* (pp. 1–19). Duke University Press.

Mirowski, P., & Plehwe, D. (Eds.). (2015). *The road from Mont Pèlerin: The making of the neoliberal thought collective, with a new preface*. Harvard University Press.

Mitchell, V., Henthorne, T. L., & George, B. (2020). Making sense of dark tourism: Typologies, motivations and future development of theory. In *Tourism, terrorism and security*. Emerald Publishing Limited.

Oakeshott, M., & O'Sullivan, L. (2014). *Michael Oakeshott: Notebooks, 1922–86* (Vol. 6). Andrews UK Limited.

Pack, S. (2006). *Tourism and dictatorship: Europe's peaceful invasion of Franco's Spain*. Springer.

Popper, K. R. (2020). *The open society and its enemies* (Vol. 119). Princeton University Press.

Ritzer, G. (2015). Prosumer capitalism. *The Sociological Quarterly, 56*(3), 413–445.

Robinson, J., & Acemoglu, R. (2012). *Why nations fail*. Crown Publishing Group.

Rushkoff, D. (2009). *Life Inc.: How the world became a corporation and how to take it back*. Random House.

Santayana, G. (1964). The philosophy of travel. *The Virginia Quarterly Review, 40*(1), 1–10.

Semmens, K. (2005). *Seeing Hitler's Germany: Tourism in the third Reich*. Springer.

Sennett, R. (1998). *The corrosion of character: The personal consequences of work in the new capitalism*. WW Norton & Company.

Sombart, W. (1913). *The Jews and modern capitalism*. TF Unwin.

Stilz, A. (2009). *The liberal loyalty: Freedom, obligation and state*. Princeton University Press.

Veblen, T. (1973). *The theory of the leisure class*. Houghton Mifflin, pp. 468–646.

Wallerstein, I. (2011). *The modern world-system IV: Centrist liberalism triumphant, 1789–1914* (Vol. 4). University of California Press.

Walton, J. K. (Ed.). (2005). *Histories of tourism: Representation, identity and conflict* (Vol. 6). Channel View Publications.

Warde, A. (2017). *Consumption: A sociological analysis*. Palgrave Macmillan.

Weber, M. (2009). *From Max Weber: Essays in sociology*. Routledge.

Wong, C. K. S., & Kwong, W. Y. Y. (2004). Outbound tourists' selection criteria for choosing all-inclusive package tours. *Tourism Management, 25*(5), 581–592.

CHAPTER 4

Tourist Gaze Reconsidered

Introduction

The theory of mobilities has gained significant recognition and traction over recent decades, illustrating mobilities' influence on daily life and the rise and expansion of globalization worldwide (Cresswell, 2011; Shaw & Hesse, 2010). Mobilities theory has become a leading paradigm within the constellations of social sciences. Anthropologists, geographers, and economists are mainly captivated by mobilities theory (Davies, 2012). As John Urry (2016) puts it, the proliferation of leisure sites such as fast food, tourist destination, and recreation parks called for a new infrastructure that is systems-oriented to improve the means of transport. The technological breakthrough of the twentieth century, in particular the proliferation of steel and petroleum cars, paved the way for the rise of a new hyper-mobile society. Historians and archaeologists who discover our civilization in a hypothetical future would think that we live in a society based on movement.

> Movement became significant in the contemporary world- indeed the freedom of movement, as represented in popular media, politics and the public sphere, is the ideology and utopia of the twenty-first century. The UN and the EU both enshrine rights to movement in their constitutions. More than knowledge, more than celebrity, more than economic success itself, it is the infinity of promised and assumed consumption of possibilities arising from

M. E. Korstanje, B. George, *Mobility and Globalization in the Aftermath of COVID-19*,
https://doi.org/10.1007/978-3-030-78845-2_4

> multiple movements that characterize the neoliberal dream. Also many people have mobility thrust upon them as the number of refugees, asylum seekers and slaves hit records of levels in the early twenty-first century. (Urry, 2016: 4)

Without any doubt, Urry revolutionized the cultural studies field, shedding light on the needs of theorizing on mobilities. To some extent, however, what would happen if mobilities theory would be an ideological paradigm that gives us a false illusion of freedom while limiting us to real immobility? What we see or, what is more important, what we need to gaze far from following a serendipity logic is finely ingrained in a cultural matrix that precedes the culture. At the time the light illuminates one sector, it obscures the opposite. Following the ideals of Enlightement logic, the tourist gaze, like the scientific gaze, throws light on space while other repressed world looms large (Sheller & Urry, 2004). The tourist gaze, as Urry and Larsen clarify, has nothing to do with the medical gaze. Rather it is finely attuned to the psychological needs of consuming to maximize pleasurable experiences. When tourists travel, they are often looking for something new, a novel or a unique experience. They are moved by the desire of gazing at anything they meet. In view of this, the tourist gaze is structured and organized in the social nature of man. It is noteworthy that lay-people gaze through a filter formed by previous stereotypes, skills, desires, and expectations demarcated by social class, gender, and nationality (Urry & Larsen, 2011).

> We elaborate on the process by which the gaze is constructed and reinforced, and consider who or what authorizes it, what its consequences are for the places which are its objects and how it interrelates with other social practices. The tourist gaze is not a matter of individual psychology but of socially patterned and learnt ways of seeing. (Urry & Larsen, 2011: 2)

Taking this as the main cue, it is reasonably clear that if production and society change, the meaning of the tourist gaze changes too. Mandkind, for some reason, witnesse periods of high and low mobiities. In its history, men have venerated or avoided travels. The feudal period, in particular, seems to be an epoch when expeditions and tours do not abound. The feudalization process, associated with the internal feudal wars, made it almost impossible for travelers to move freely and safely. However, as we have already discussed, some prosperous civilizations and empires cultivated travels and expeditions. The imperial structures constructed paths and ways to allow a secure and easier exchange of goods while mobilizing the army machine in case of riots or revolts. Hence, leisure travels and empires were historically intertwined. After the turn of the century, the high-mobility society has gone through a deep fracture, where tourists are not welcomed any longer. The attacks on the World Trade Center humiliated the US and showed Western civilization how modern leisure becomes a weapon. On September 11, 2001, four commercial airplanes were seized and weaponized against the symbols of military machine and commercial power of one of the world's most important countries (Graham, 2001; Sheller & Urry, 2004). These blows changed the spirit of the Occident as well as the ways the "non-Western Other" is portrayed. Not surprisingly, the West closed its doors to the "Otherness" by introducing more restrictive surveillance and control protocols, even mining the powers of checks and balances characteristic of democracies (Barber, 2004; Wilkinson, 2011). The stock market crisis of 2008 made being open to "Others" even more problematic. This financial crisis had consequences that impacted in two directions. On the one hand, globalization set the pace for more feudalized regional economies, which cemented the authorities of populist leaders (Campbell & Hall, 2017; Berman & Snegovaya, 2019). In other cases, some old regional separatist discourses suddenly surfaced to set a political agenda in Europe (see the case of Barcelona) or northern Italy without mentioning Francophone Canada (Bosco & Verney, 2012; Khenkin, 2018). Doubtless, without the culture of terror inspired by terrorism and the crisis of 2008, Donald Trump would never have reached the presidency of the US (Altheide, 2017). On the other hand, the economic stagnation embodied a process of demonization around the "foreign tourist" (Korstanje & George, 2020).

Some discourses aligned with the needs of protecting the domestic industry correspond with national chauvinist narratives that articulated putting the nation first (Giroux, 2017; Nail, 2016). It was time to see slogans like America first! and Make America great again! Similar nationalism can be found in Germany, Spain, France, and Belgium. Terrorism earlier and the crisis later fractured the illusory doom of mobilities. Still further, the resulting budget cuts because of the financial crisis in combination with the mass migration of Muslim countries, most of them devasted by civil wars, accelerated an economic paralysis in the developing economies of the West. Despite the military invasion of the US in the Middle East, more radicalized cells gradually were taking a territorial position of some cities in Syria and Iraq, forming an Islamic political unity known as the Islamic state ISIS. Urban and tourist cities such as Paris, London, and Brussels have been the targets of terrorist attacks since 2015. If the death of Aylan Kurdi, a five-year-old Syrian boy who was found dead on the Turkey coast, shocked the Western social imaginary which pressed for the governments to increase the migration quotas, these attacks closed the doors of Eden (Nail, 2016) again. Terrorism gradually undermines the Western civilization from within, beginning not only a new identity crisis but also the decline of hospitality. Today, different events mark the end of hospitality, ranging from Islamophobia, Brexit, tourist-phobia, and the Windrush scandal to even promises of building a wall in the US-Mexico borders (Korstanje, 2017, 2018; Matusitz, 2020). In the streets of Barcelona, locals make graffiti that say "tourists you are the real terrorists!" Of course, COVID-19 seems not to be the exception to the rule. In days of this global pandemic, we have coined the term *wicked gaze* to denote a new facet of the capitalist world where the tourist is no longer a desired guest. This point will be continued in later chapters. In the new normality, "the Other" is presented as an intruder parasite that potentially infects society with a lethal virus. Since nobody knows who is infected, all citizens are considered potential carriers of the virus and forced to live in isolation.

This chapter reviews the strengths and weaknesses of the mobilities paradigm for constructing a bridge between Marxism and the cultural theory. In so doing, we place the valuable contributions of Urry and mobilities theory under the critical lens of scrutiny. The term *tourist gaze* speaks to us of a socio-cultural matrix that precedes tourists as well as their cosmologies. In view of this, capitalism moves and consolidates in a net of exchanges which crystallize global reciprocity, or as Urry says "economies of signs and spaces". However, one of the aspects Urry did not take into consideration was the figure of *hospitality*. Like Marxism, the cultural theory has no minimum efforts to deal with hospitality as a social institution that describes Western civilization's architecture. In Chaps. 6 and 7, we shall carefully review the anthropological description of hospitality and its interconnection with the global capitalist system, where the "Other" catalyzes pleasure maximization while cancelling any possibility of genuine contact. In this chapter, we assess the notion of the tourist gaze as well as the recent advances in mobilities theory. In the first section, we place Urry's legacy under the microscope. His notion of the tourist gaze resonated heavily in social sciences and tourism research and is useful for understanding the post-COVID-19 days. Urry opened the doors of materialist historicism to create a novel theory about capitalism and mobilities. Our tourist experiences depend upon the economic forces of society as well as the organization of human relations. In the second section, we lay the foundation for a new theory of mobilities. We hold the strong opinion that far from living in a mobile world only a small percentage of the population has the right to move freely. The creation of the modern nation-state was based on a false sense of mobility that was functional to countless nomad ethnicities' domestication.

John Urry and the Tourist Gaze

John Richard Urry was a British sociologist who does not need an introduction. He was born on June 1, 1946, in London, England, and died on March 18, 2016, in Lancaster. As a full professor at Lancaster University, he contributed notably to mobilities and cultural theories. Based on transactional environmentalism, he toyed with the belief that capitalism evolved to more abstract forms changing the means of production and social interaction (Urry, 2001, 1992; Urry & Larsen, 2011). In his *Anatomist of Capitalist Societies*, he exerts an eloquent critique argumentation on Marxism's conceptual limitations. He decided to overcome the obstacles of classic Marxian texts originally demarcated by Althusser. Although Urry believes Marx deserves recognition for his materialist view of history, he retains civil society's idea to describe those institutionalized practices, customs, and values that explain our social behavior. Actors constitute themselves as the real subject when they enter civil society, as Urry (1981) contends. In consonance with this, the notion of class, at least as Marxists imagine it, has no direct relation to ideology but to civil society as well as the negotiation of the agency with the structure. In 1987 he published jointly with Scott Lash an interesting book entitled *The End of Organized Capitalism*. Urry and Lash devoted their efforts toward constructing a bridge between Max Weber and Karl Marx but were giving a fresh insight. The founding fathers of sociology wrote at a time when capitalism was in its acme. This era is now coming to an end. In this vein, Lash and Urry (1987) acknowledge that mass production is setting the pace for the new disorganized capitalism where the relationship among classes, spaces, and culture is blurring. It is vital to say that capital—far from being controlled—expands beyond the sovereignty of the nation-state. Industries, banks, and corporations are investing outside to avoid the financial controls of nation-states. This process accelerates a movement toward economic decentralization, which seems to be the precondition for globalization.

Along with Lash, Urry again released in 1994 what is probably one of his masterworks, *The Economies of Signs and Spaces.* Lash and Urry realized that there are substantial changes in the lines of global geographies, which correlate directly with a much deeper decentralization of economic production. On the question of whether the tourist gaze organizes the economic forces of production into a coherent cultural matrix, it is important to think of that gaze within the hegemony of cultural relations. Far from being a disorganized practice, the tourist gaze engages directly with a cultural matrix that tells us what can be gazed at or not. Our current obsession with outstanding experiences is based on that matrix; the fact is that Lash and Urry understand that classic tourism is mutating to a global form of consumption. Lay-people travel long distances to be in contact with exotic cultures. The capitalist world is recreating the conditions for the passage from a culture mainly centered on low mobilities to that centered high mobilities. Having said this, the trajectory of goods, bodies, and cultures has created a climate of exchange where social ties are undermined. They use Maussian gift theory to illustrate the ways global trade affects social relations. Free trade leads economies to produce abstract commodities whose values are conferred by the aesthetics of signs. Such an exchange is baptized by Lash and Urry as an *aesthetic reflexibility*. In the Occident, especially in the nineteenth and twentieth centuries, the old reflexibility was replaced by a new, more romantic one oriented to an idealized (but simplified) view of the "Other". Travelers began not only to colonize and classify the outside world but also to look for unique experiences. Once the capitalist system successfully expanded, travelers became commodities to be gazed at (consumed) by other travelers. Similarly, the net of experts acted as a source of information that remapped the world, dividing the safe from the unsafe destinations. Travels were regulated according to tourist security. The wild zones, which are characteristically risky destinations, were avoided while safe tourism was recommended by the travel experts (tour operators). The destination, not the "Other", was preferably the stepping stone in the architecture of modern tourism (Lash & Urry, 1993). Today, the tourist gaze mediates between citizens and social institutions, a point which led to a discussion with Dean MacCannell (2001). While for MacCannell the structure determines the social action, for the case of the tourist behavior, Urry goes on to admit that each agency, within the constellations of possible results offered by the cultural matrix, adopts, negotiates, and develops its social practices. But what is the tourist gaze?

It is important to note that Urry borrows the term *gaze* from the Foucauldian insights based on the medical gaze. For Foucault, the medical gaze is explained by introducing Western rationality, which looks for controlling things through gazing. The power of science emanates from its authority to gaze while elaborating a diagnosis of the problem (sickness). The West has fleshed out an uncanny fascination for ocularcentrism. This motivates Urry to extirpate Foucault's term to define the tourist gaze as the complex interplay of signs, discourses, allegories, and practices of seeing where the gazed object or subject is controlled and possessed. Hence the act of gaze is subordinated but at the same time shaped by a cultural matrix that marks what is desirable or undesirable. Understanding how this matrix works seems to be the most efficient method to describe how society functions without mentioning the relationship between the different gazers. To wit, Urry (2002) distinguishes three types of gazes—romantic, individual, and solitary. As he remarked, "I call the romantic gaze, solicitude, privacy and a personal, semi-spiritual relationship with the object of the gaze are emphasized. In such a case, tourists expect to look at the object privately or at least only with significant others" (Urry, 2002: 150).

Devastated or desert areas that in normal conditions inspire terror are revitalized and re-organized into commoditized landscapes while often connecting to *an imperial economy*. The distance between gazers and the gazed object plays an important role in cementing the imperial authority. In this token, ocularcentrism is the key factor that explains the European expansion in the earlier centuries, at least for Urry. A closer look, in theory, shows that the cultural values of modern societies bestow on their citizens a global right to travel everywhere, touching different cities and dispersing geographical cosmologies. These rights are unnatural and consecrated in the economic division of labor typical of global capitalism. The connection between leisure and labor explains that citizens work hard the entire year to spend their amassed wealth on holidays. Tourism is defined as a mechanism of escapement where the self abandons unique experiences. While tourism is open to the logic of escapement, leisure practices legitimate the liberalization of all social constraints. But what happens if the political authority of institutions declines?

Urry goes on to say that tourism consumption—likely in consonance with MacCannell—fills the gap mediating between lay-citizens and their authorities. The act of gazing is always an imperial practice that aims to possess the gazed one. As Korstanje and Seraphin (2017) wrote:

> Gazing is a way of possessing what is being gazed but he said, this can be only done in a hypermobile context, where cultures and their peoples are marked by the sign of central economies. The process of commoditization proposed by tourism can be materialized only because objects are conferred of a sign, which is exchanged in the market. We do not consume products by their features as a couple of centuries ago, but by the imposed sign over them. Coffee from Colombia, Whisky from Scotland, Wine from Argentina are only some examples that validate Urry's worries. (p. 8)

Most certainly, Urry departs from the premise that the current high mobility derives from the dominance of aesthetic modernity where the needs of gazing surpass other senses. We live in a society of spectacle—echoing Debord—where the tourist travels embody the archetypes of happiness, development, and emotional engagement. From this stance, the nation-states are re-inventing their geographies, borders, and identities constantly confronted by a counter-force: the mass migration. Finally, the current postmodern means of production has formed a bubble of illusion that does not permeate with reality. To some extent, the capitalist society has serious problems in reversing the contamination that accelerates climate change. Urry contends that *economist theorists* who struggle to reverse the negative effects of climate change misjudge the tenets of carbon capitalism which is based on a commoditization process. These economists suggest that energy may very well be replaced by oil but *carbon capitalism* has not been designed to function with energies, only with commodities. This assumption prompts Urry to assert that to change the current ecological crisis, society should adopt other mainstream cultural values (Urry, 2009, 2010, 2015). Sociologists are boldly urged to imagine new methods and mobilities paradigms to diagnose global capitalism's social maladies (Birtchnell & Caletrío, 2013; Sheller & Urry, 2016) or the contradictions of mobilities theory (Korstanje, 2018). Considering his prolific view and excellence in research, we do apologize to colleagues who are experts in Urry's studies for this simplification. We attempted to highlight only a snapshot of Urry' legacy so that readers who are unfamiliar with his writings digest the main argumentation. The figure of tourist gaze not only explains the psychological motivations of modern tourists but also grounds a fertile foundation that helps to understand the hard days we are living in. Urry wrote at a time when the capitalist ethos was in expanse, and he has never seen the aftermath of COVID-19.

Our current critique of Urry's texts is moved by his lack of interest in describing the connection of mobilities with hospitality. Urry was never interested in inspecting either history or the past of a premodern Europe. The theme will be approached in the following section. After all, do we live in a mobile society?

Contributions and Contradictions of Mobilities Theory

Exegetes and supporters of John Urry claim that even if we live in a mobile world some incompatible cosmologies contrast notably with it. They do not deny the hapless presence of asylum-seekers or people who are trafficked or pressed to live in immobility (social mourning) (Hyndman, 2012). But they do believe that these are asymmetries or maladies generated by a new form of decentralized capitalism, which started with the end of the welfare state (Hannam et al., 2006; Adey & Bissell, 2010). It is important to mention that Urry has shed light in many sub-fields of sociology ranging from terrorism (Diken & Laustsen, 2002; Diken, 2004; Amoore, 2006; Tzanelli & Korstanje, 2019), climate change (Parsons, 2019; Suliman et al., 2019), techniques of surveillance (Pickering & Weber, 2006), and consumption to dark neocolonial exploitations (Tzanelli, 2015, 2016, 2018) without mentioning the creation of a new epistemology to set the research agenda (Vannini, 2010; Jensen et al., 2020).

For supporters of mobilities theory, immobility should be studied as a consequence, not the root, of capitalist exploitation. The productive capitalist system began in the colonial period to colonize the world, creating some material inequalities. While capitalism expanded, some contradictions arose. Capitalist production was made stronger in two directions. On the one hand, it stimulated goods exchange, transferring capital from the poorer to richer nations. On the other, the migrant workforce which struggled to enter rich cities was undermined as "undesired citizens" (Martell, 2017). The same applied in the case of automobiles, as Urry overly proclaimed. Urry certainly acknowledged that mobilities opened the doors to technical expansion but at a higher cost. The automobile industry grew exponentially every year, replacing other classic means of transport such as the train and the horse. Conceived in this way, automobility is a system that coerces lay-people *into intense flexibility*. The opposite is equally true; automobility helps drivers to move elsewhere but limited to a net of roads and closed ways. Urry called automobility a Frankenstein-created monster that emancipates while dominating drivers (Urry, 2004). Of course, this stance begs two significant questions: how mobile are we? And are modern mobilities a genuine form of displacement?

To these questions, Urry would add the point that the advancement of modern capitalism accelerated some radical shifts not only in the means of production but in social behavior. Modern capitalism disposed of economic exploitation while offering tourism consumption to workers. This is why mobilities and the tourist gaze are inextricably intertwined. Mobilities exhibit a tendency oriented toward the multiplication of travels year to year followed by the shortening of the time of travels. Faster machines successfully transport passengers and goods from one point to another in hours. This double dynamic, needless to say, has some unexpected shortcomings. Like poverty, material inter-class asymmetries exposed in Urry's text are part of hyper-mobile cultures' social maladies (Urry & Larsen, 2011). To put it bluntly, immobility seems to be not only a reality of the social system but an unintended implication of capitalism (Urry, 2001). Rather, our intellectual position discusses atmosphere immobility not as collateral damage but as the ideological core of the capitalist expansion. To cut a long story short, the illusion of a false sense of mobility is endorsed to imposition a disciplinary regime where the body is an inert object of energy provision for the capital machinery (Korstanje, 2018). The best example of this hypothesis takes place in the plot of Matrix (1999), an American science fiction saga directed by the Wachowskis. In this dystopian future, humans are subdued and immersed in an unreal (virtual) reality dominated by the Matrix, an artificial superintelligence that commands intelligent machines in the conquest of the planet. A group of insurgents who refused to live in this virtual reality work with Neo to uncover the truth. In the early twenty-first century mankind is enslaved by machines whose access to solar energy is blocked by humans. In retaliation, the machines capture human bodies and extract their energies while artificially shaping their minds into thinking they are in a peaceful and comfortable world. Lying in pods, humans remain immobile, living in a simulated reality, while their bioelectric power is extracted and packaged to feed the Matrix. Another film known as *The Island* (2005) is illustrative of the same phenomenon. Kept aside in an isolated compound, Lincoln Six Echo (McGregor) and Jordan Two Delta (Johansson) have a safe and routine life, at least until they discover that they both are clones harvested to supply organs to the elite. After Lincoln unearths the dark truth that they are clones for organ harvesting as well as surrogates for wealthy citizens he convinces Jordan to escape to the outside world. There is no conflict in the compound and harmony prevails among inhabitants through the imposition of strict norms. Dr. Merrick, who commands the compound, keeps the clones domesticated. They have been told that the world has become too contaminated to support life, with the exemption of an island free of the pathogen. Each week, one resident wins the right to leave the compound and settle on the island. This paradisiacal island is the supreme goal that keeps residents subdued. While visiting another sector, Lincoln discovers that beyond the tourist travel lies a perverse system designated to exchange organs, to cultivate clones to serve an elite set of people who refuse to die. The winner, at best, is used for organ harvesting and surrogate motherhood for their wealthy sponsors, of whom they are clones.

Both plots show one of the contradictions of mobilities theory as well as the illusion of living, neglecting the tragedy as a part of the natural world. In fact, the conquest of the Americas and the expansion of the European order in the new world disciplined countless aboriginal nomad groups by not only silencing their voices and prohibiting their itineraries but forcing them to live within the margin of the nation-state. A false sense of movement was unilaterally imposed through the articulation of safe paths. In doing so, new citizens were subject to a climate of immobility and oppression (generation after generation). Domesticated to believe they were moving when they did not, citizens were re-educated in the light of leisure travels and the law of the state. The capital is the gateway to access this wonderland of mobilities and travels. Those who do not have enough money to travel are doomed to live with nothing. As Bauman suggests, while some privileged tourists are invited to traverse across the world, forced migrants are immobilized (Bauman, 2017). Like the above-discussed plots, the capitalist system employs a stick and carrot method, alternating fear with pleasure. The state of order, which the capitalist society builds, subverts when a disaster, such as COVID-19, strikes. Here is the second limitation of the theory: the term *mobilities* explains everything and at the same time nothing. We start from the premise that the world is hyper-connected where the act of traveling is a basic right widely conferred to all free citizens. However, only 20% of people travel while the remaining 80% live in a peripheral condition. In 2018, of the world's total population of almost 8 billion people, only 1.2 billon traveled for tourism or business purposes according to statistics reported by the UN World Tourism Organization (Pearce, 2019). Far from situating as a liberating force that regulates human reciprocity, as Urry stated, mobility seems to be reserved for an elite geographically located in the northern hemisphere. Furthermore, in *Mobilities Paradox*, Korstanje (2018) documented Mary and Roger, two elderly people who bought a car to move. The so-called right of mobility precipitated serious imbalances in the budget because of emerging inflation. Mary and Roger were forced to sell the car and with that their dream of mobility. This story, although abridged, reflects the problems of mobilities in the Global South. What works for some cultures seems not to work for others. Last but not least, the tourism industry and leisure travels should be traced back to the birth of industrialism, according to Urry's insights. This means that tourism would never exist beyond the zenith of modernism.

In a well-read paper published in the prestigious British journal *Theory, Culture & Society*, Urry (1988) acknowledges that tourism is a vital force for industrialism simply because it represents leisure development as integrally bound to "paid-work". Starting from the premise that paid work is the precondition for tourism, Urry adheres to the theory that tourism in popular seaside resorts is changing to "contemporary tourism", which expresses the rise of a *post-industrial society*. What is equally important, his main thesis is that if the classic tourist experience born in the nineteenth century was reached at the time the site was visited, the postmodern tourists replicated countless experiences by articulating fabricated scenes that are replicated by digital technology again and again. In this vein, Urry goes on to write:

> The post-tourist is freed from the constraints of simply high culture on the one hand or the untrammelled pursuit of the pleasure principle on the other. He or she can move easily from one to the other and indeed to gain pleasure from the connection between the two. The world is a stage and the post-tourist can delight in the multitude of games that can be played. When the miniature replica of the Eiffel Tower is purchased, it can be simultaneously enjoyed as a piece of kitsch, an exercise in geometric formalism, and as a socially revealing artefact. (Urry, 1988: 38)

Like MacCannell, Urry strongly believes tourism to be the maiden of modernity. By adopting this conception, he does not appreciate the intersection of travels and empires. As Riesman brilliantly documented, empires regularly go from an inner to an other-directed character. The civilized order is not a linear facet of history or the history of production but a cyclical balance of forces and counter-forces. All empires have invented their version of the "Otherness" accompanied by an ideological justification of travels. Modern tourism, as Urry carefully contemplates, coincides with the expansion of the Anglo-American empire. Modern mobilities can be equated to a carousel that moves in circles centered on its axis. Why have mobilities theory gained so much traction and recognition in recent years?

Geoffrey Skoll (2014) gives a potential answer. He sets forward a fresh argument that explains society debates around dialectics between the agency and the collective thought. Beyond the society, the agent has no significance, while the society loses its reason to exist without human agency. The dialectical negotiation between the two parts needs a third one which is naturalized. The role of consciousness in Freud equates to capital in Marx. Both are social constructs unfolded to describe a dialectical relation, but in doing so they disappear under the waves like anvils. Behind its self-explicatory nature, the term avoids being marked and correctly deciphered. Unfortunately, Urry falls into the same philosophical dilemma. By assuming the world to be hyper-mobile because of the overproduction of cars, the growth of the tourism industry, and several international flights without noting the digital technological revolution that has occurred in recent years, we involuntarily confirm our hyper-mobile hypothesis. In this game, we run the risk of placing the cart before the horse.

To be just with Urry, we have to say that he had left an impressive legacy to be followed by theorists of mobilities. He created a positive revolution in the field of social sciences, first and foremost at a moment when mobilities occupied a marginal position in sociology. The turn of the century evinces the proliferation of many unseen risks. Countless natural disasters, virus outbreaks like COVID-19, and terrorism rattle the notion of peace and order of global capitalism. The sense of flexibility given by the democracy to the Western tourists is accompanied by counter-forces disposed around tourism security (Hannam et al., 2006). Democracy is the centerpiece here that delineates the borders of high- and low-mobility societies. Toying with the belief that mobility is a partial project, so to say successfully implemented in the West but falling short in other non-Western cultures, Hannam, Sheller, and Urry conceive that mobilities theory has the potential to be applied to a vast specter of themes in the years to come. In fact, in the era of digital mobility, tourism is being shaped into virtual forms. By the same token, Haddington et al. (2013) alert how the adoption of new digital technologies applicable to tourism is engendering new forms of displacement where the paradigm of being there or being mobile is no longer important. They contend that psychical displacement is not the decisive criterion to define mobilities theory today. Regardless of their goals, technology ascribes to specific forms of control and socialization.

Toward a New Conception of Mobilities

As Gale (2009) highlighted, the modern tourism industry experiences a set of disorganized forms of consumption that harm mobilities' sacred aura. Gale is concerned about what he dubbed as the end of tourism, as least as we know it. Virtual tourism affirms a new pleasurable experience but blurs the borders between work and pleasure. In accordance with this, Kevin Hannam (2009) urges redefining the epistemological basis of mobilities theory because of global capitalism's technical advances. In reply to Gale, Hannam enthusiastically argues that tourism is not in decline but is in a moment of radical transformation. As he adds, sociology should adapt its methods and applied research to new horizons, leaving the old one behind. Since any place is far from being an objective construct simply because it is subject to cultural dynamics which change in time, the sense of mobilities operates in the dense net of various modes of representations often stemming from relational moments emotionally aroused or evoked.

In the post-COVID-19 society, people do not need movement to make tours. Virtual tourism—supported by the revolution in digital technologies—situates as one of the most consumed services during the lockdown. In fact, this leads us to rethink the paradigm of mobilities theory in a society without travels or re-feudalized according to new territorial borders (Molz, 2020). Virtual tourism is probed to be a successful instrument of economic recovery in contexts of strict lockdown because of COVID-19. Simultaneously, while the pandemic exposed the vulnerabilities of tourism system technologies offer an interesting toolkit toward new forms of tourism based on external reality and virtual consumption (Mizrachi & Gretzel,

2020; Kwok & Koh, 2020; Stankov & Gretzel, 2020). Ulrike Gretzel et al. (2020) pay attention to the role played by e-tourism toward a transformative research agenda. The new beginning that COVID-19 has initiated gives further plurality and transparency to tourism research. As they note,

> The profundity and complexity of the impact of COVID-19 require both a short-term reaction and a long-term readiness in order to understand some of its far-reaching effects at the fundamental level. As the intersection between IT and tourism, the field of e-Tourism must now grapple with the two "fronts" of a fast-changing new reality and tremendous prospects for long-term change. As e-Tourism researchers, we find this new reality provides both enormous challenges and exciting opportunities from the perspectives of scientific inquiry and technological innovation. e-Tourism research must now elevate itself to be ready to ask, and prepare to find answers to, some critical and significant questions regarding the role of IT in contributing to a better future of tourism and society. (Gretzel et al., 2020: 2)

Anyway, COVID-19 is causing some worries in authorities regarding ethnic violence and the potential geopolitical disputes among nations. What is more important, there are no clear indicators on whether technology helps reduce or gravitate to xenophobic expressions. In this vein, Mostafanezhad et al. (2020) alert that the global pandemic is now sharpening long-dormant prejudices and racist discourses toward Asian tourists. In view of this, racial discourses epitomize *orientalist assumptions* dotted of political mistrust and fear of the "non-Western lifestyle". Tourism takes shape in the host-guest encounter which activates some natural anxieties because nobody knows the intention of the "Other". Still, now these anxieties are being challenged to the logic of hate and resentment. Paradoxically, in a world without tourists, tourists are undesired persons.

> Within the arsenal of geopolitical blame game contestants, the structural drivers and consequences of the fallout from the COVID-19 pandemic are obscured by historically rooted, xenophobic narratives that are weaponized through touristic imaginaries of place. Navigating the uncertain future of global (im)mobility, scholars may consider how tourism encounters both challenge and reinforce geopolitical narratives, and by what means tourism geo-economics reckons with its recently exposed wounds of structural inequality. Resuming travel will depend on the lifting of border restrictions

> and whether countries see each other as 'safe enough' to allow the flow of people-vectors between them. While highly contingent, geopolitical relations will underline reopening efforts. (Mostafanezhad et al., 2020: 185)

We have reviewed the importance of and legacy left by John Urry, who is considered one of the fathers of mobilities theory. The recent disrupting events have anyway altered the ways and morphologies of leisure travels. Urry helps understand these new times and gives a fresh insight to place the durable effects of COVID-19 in perspective. The future of mobilities theory is uncertain, but what is clear is that the time is ripe for mankind to reconfigure geopolitical relationships. The global ethos has been mutated to a decentralized new world similar to the feudalization process initiated after the Roman Empire collapsed. Without any doubt, the war on terror makes way to the war against a virus.

The War Against a Virus? Conclusion

Like COVID-19 that comes and goes back in waves, imperialism expanded through three clear-cut waves. Has the war on terror mutated and calibrated the target, now toward a virus? To write about COVID-19 seems to be a difficult task given the velocity of media events. The virus originated in the Chinese province of Wuhan toward the end of 2019, and of course, it was rapidly disseminated worldwide. Despite the attempts of the Chinese government to exert censorship, Western media and press covered the pandemic 24/7 in real time. To wit, experts agree that for each case diagnosed other ten potential cases will remain silent without or with mild symptoms. COVID-19 terrorized the world, which prepared to fight against an invisible foe. To put things in perspective, COVID-19 shocked the world and paralyzed the global trade, leading many service sub-sectors to bankruptcy. To stop the pandemic, governments adopted different restrictive measures which oscillated from strict lockdowns to the closure of borders, airspaces, schools, and shopping malls. Tourists stranded were systematically repatriated while hotels were refurbished as hospitals. Tourists, who were seen as ambassadors of the prosperous capitalism, were suspiciously labeled as potential mortal disease carriers. In Urry's terms, the tourist gaze became a wicked gaze.

As already discussed, the first wave of modern imperialism started with the conquest of the Americas. If the eighteenth and nineteenth centuries witnessed how the Western civilization expanded to colonize the world, inventing the figure of the "non-Western Other", aborigines represented a priority for European ethnographers who began to document their cultures before their disappearances. The modern nation-state disciplined the "non-Western Other" forcing him to live within its margins. The quest for novel experiences with the "Other" motivated European travelers to go beyond Western civilization's borders. The notion of being there was important to understand why many Europeans looked at travel as a sign of status and distinction. These travels inspired writers to create a new literary genre (travel writings) and cemented the authority of science worldwide. The tourism industry, one might speculate, rests on this colonial logic whereby "the Other" is a cosmocreator of a personal experience. In the threshold of time, the native population not only scrambled for further rights but adopted democracy as an ideal model. This inter-ethnic bridge served as the illusion of a dialogue between the former imperial powers and their peripheries. The notion of "being secure here" or "being in an insecure there" radically shifted, and the world was finally flattened: no borders, no difference introduced a crisis for ethnographic knowledge. As debated in earlier chapters, it is safe to say that an ethnocentric version of ancient hospitality was politically adapted to the imperial goals. The second wave of imperialism mutated just after the attacks on the World Trade Center in 2001. This death blow negatively affected global capitalism. Four commercial airplanes were seized and weaponized against commercial and military targets of the US. From that moment on, the West interrogated further its endorsement of mobilities, debating the risks of destroying the geographical frontiers anew. The 9/11 attacks, as a foundational event, caused serious changes for the Western democratic institutions, accelerating the decline of Western hospitality. The psychological fear instilled in the US and Europe activated more aggressive protocols of surveillance, including border closures, the advocation of supremacist discourses, Islamophobia, and tourist-phobia. Tourists as imagined non-Western "Others" are not admired any more, but they are feared, hated, and considered undesired guests.

The stock market crisis that happened in 2008 affirmed such a tendency. Some reactionary circles took advantage of the crisis, blaming globalization and some ethnic minorities (as migrants and asylum-seekers) as potential terrorists. Jair Bolsonaro, Donald Trump, and Viktor Orban won under the auspices of economic stagnation. Terrorism engineered a new movement that does not see the enemy as coming from the outside but living within—probably like an evildoer who looks like us. Such a metaphor determined a culture of fear that eroded human reciprocity as well as social ties. In this postmodern cosmology, a terrorist is now a person who looks or behaves like us; he or she might even be a native of the targeted society. If during the nineteenth-century colonial rule, Westerners consumed the alterity of colonized Others, in post-9/11 contexts, they learned to cannibalize their essence, by blurring the boundaries between the risks of being here and being there.

What Has Finally Changed with the COVID-19 Outbreak?

COVID-19 is the third blow to the imperial order, an event that affirms a tendency to self-cannibalization. Like the terrorist who lives within, the virus, which circulates everywhere, is hosted by our body—for anyone can be turned on COVID-19's behalf into a lethal weapon. This brings us to the metaphorical extensions of the "quarantine", as the radical neglect of the neighbor. Citizens are confined, so they live in isolation, like war prisoners, so that the infection of other citizens is prevented. As silent carriers of the public enemy par excellence, who display no symptoms of infection exteriorly, we are considered potential terrorists. Those who violate the quarantine are repressed, or even jailed as an attack on public health. Like the 1970s war on cancer, the war against criminality in the 1980s, or the war on terror in the twenty-first century, we are now caught in a tug of war on a virus. These SARS-family viruses do decimate animals but are mainly designed to strengthen the population gene pool. Although easily reproducible viruses of high mortality are uncontrolled, we argue that this virus is potent because it allows for policing individual behavior (e.g., what people do or do not do to prevent its spread). The tourist industry and the war on terror are inevitably entwined; the viral crisis ended with the tourist gaze recycling of the Western rationale while the aesthetics of tourism consumption changed. Hotels will be recycled as hospitals and airplanes will be placed at the disposal of governments to take home expatriates and stranded tourists. Here, there is a clear retransformation of modern tourism's conditional hospitality—à la Derrida—into the unconditional hospitality provided by the hospital. COVID-19 has strengthened the precautionary logic of modern polities after 9/11. It has inverted the old logic of mobilities by suggesting the end of conditioned hospitality that we associate with tourism. This much deeper issue will be re-visited in the last chapter.

References

Adey, P., & Bissell, D. (2010). Mobilities, meetings, and futures: An interview with John Urry. *Environment and Planning D: Society and Space, 28*(1), 1–16.

Altheide, D. (2017). *Terrorism and the politics of fear*. Rowman & Littlefield.

Amoore, L. (2006). Biometric borders: Governing mobilities in the war on terror. *Political Geography, 25*(3), 336–351.

Barber, B. R. (2004). *Fear's empire: War, terrorism, and democracy*. WW Norton & Company.

Bauman, Z. (2017). Tourists and vagabonds: Or, living in postmodern times. In *Identity and social change* (pp. 19–32). Routledge.

Berman, S., & Snegovaya, M. (2019). Populism and the decline of social democracy. *Journal of Democracy, 30*(3), 5–19.

Birtchnell, T., & Caletrío, J. (Eds.). (2013). *Elite mobilities*. Routledge.

Bosco, A., & Verney, S. (2012). Electoral epidemic: The political cost of economic crisis in Southern Europe, 2010–11. *South European Society and Politics, 17*(2), 129–154.

Campbell, J. L., & Hall, J. A. (2017). *The paradox of vulnerability: States, nationalism, and the financial crisis*. Princeton University Press.

Cresswell, T. (2011). Mobilities I: Catching up. *Progress in Human Geography, 35*(4), 550–558.

Davies, A. R. (2012). Geography and the matter of waste mobilities. *Transactions of the Institute of British Geographers, 37*(2), 191–196.

Diken, B. (2004). From refugee camps to gated communities: Biopolitics and the end of the city. *Citizenship Studies, 8*(1), 83–106.

Diken, B., & Laustsen, C. B. (2002). Zones of indistinction: Security, terror, and bare life. *Space and Culture, 5*(3), 290–307.

Gale, T. (2009). Urban beaches, virtual worlds and 'the end of tourism'. *Mobilities, 4*(1), 119–138.

Giroux, H. A. (2017). White nationalism, armed culture and state violence in the age of Donald Trump. *Philosophy & Social Criticism, 43*(9), 887–910.

Graham, S. (2001). In a moment: On glocal mobilities and the terrorised city. *City, 5*(3), 411–415.

Gretzel, U., Fuchs, M., Baggio, R., Hoepken, W., Law, R., Neidhardt, J., ... & Xiang, Z. (2020). e-Tourism beyond COVID-19: A call for transformative research. *Information Technology & Tourism, 22*(2), 187–203.

Haddington, P., Mondada, L., & Neville, M. (2013). *Interaction and mobility, language and the body in motion*. De Gruyter.

Hannam, K. (2009). The end of tourism? Nomadology and the mobilities paradigm. In J. Tribe (Ed.), *Philosophical issues in tourism* (pp. 55–70). Channel View Publications.

Hannam, K., Sheller, M., & Urry, J. (2006). Mobilities, immobilities and moorings. *Mobilities, 1*(1), 1–22.
Hyndman, J. (2012). The geopolitics of migration and mobility. *Geopolitics, 17*(2), 243–255.
Jensen, O. B., Bueno, A. V. H., Smith, S., & Christensen, C. B. (2020). Methods of mobilities design research. In *Handbook of research methods and applications for mobilities*. Edward Elgar Publishing.
Khenkin, S. M. (2018). Separatism against the state: The case of Catalonia. *Polis. Political Studies, 3*(3), 166–175.
Korstanje, M. E. (2017). *Terrorism, tourism and the end of hospitality in the West.* Palgrave Macmillan.
Korstanje, M. E. (2018). *The mobilities paradox: A critical analysis.* Edward Elgar.
Korstanje, M. E., & George, B. P. (2020). Education as a strategy to tackle over tourism for overtourism and inclusive sustainability in the twenty-first century. In H. Seraphin, T. Gladkikth, & T. T. Vo (Eds.), *Overtourism* (pp. 341–359). Palgrave Macmillan.
Korstanje, M. E., & Seraphin, H. (2017). Revisiting the sociology of consumption in tourism. Chapter 2. In S. Kumar Dixit (Ed.), *Routledge handbook of consumer behaviour in tourism and hospitality* (pp. 2–16). Routledge.
Kwok, A. O., & Koh, S. G. (2020). COVID-19 and extended reality (XR). *Current Issues in Tourism, 24*(14), 1935–1940.
Lash, S., & Urry, J. (1987). *The end of organized capitalism.* University of Wisconsin Press.
Lash, S. M., & Urry, J. (1993). *Economies of signs and space* (Vol. 26). Sage.
MacCannell, D. (2001). Tourist agency. *Tourist Studies, 1*, 23–37.
Martell, L. (2017). *The sociology of globalization.* Polity Press.
Matusitz, J. (2020). Jihad: A description. In *Global Jihad in Muslim and non-Muslim contexts* (pp. 29–57). Palgrave Macmillan.
Mizrachi, I., & Gretzel, U. (2020). Collaborating against COVID-19: Bridging travel and travel tech. *Information Technology & Tourism, 22*(4), 1–8.
Molz, J. G. (2020). MoVE: Mobile virtual ethnography. In M. Buscher et al. (Eds.), *Handbook of research methods and applications for mobilities* (pp. 193–201). Edward Elgar Publishing.
Mostafanezhad, M., Cheer, J. M., & Sin, H. L. (2020). Geopolitical anxieties of tourism: (Im)mobilities of the COVID-19 pandemic. *Dialogues in Human Geography, 10*(2), 182–186.
Nail, T. (2016). A tale of two crises: Migration and terrorism after the Paris attacks. *Studies in Ethnicity and Nationalism, 16*(1), 158–167.
Parsons, L. (2019). Structuring the emotional landscape of climate change migration: Towards climate mobilities in geography. *Progress in Human Geography, 43*(4), 670–690.

Pearce, P. L. (2019). Are tourists interesting? In P. Pearce (Ed.), *Tourist behaviour* (pp. 2–19). Edward Elgar Publishing.

Pickering, S., & Weber, L. (2006). Borders, mobility and technologies of control. In *Borders, mobility and technologies of control* (pp. 1–19). Springer.

Shaw, J., & Hesse, M. (2010). Transport, geography and the 'new' mobilities. *Transactions of the Institute of British Geographers, 35*(3), 305–312.

Sheller, M., & Urry, J. (2004). *Tourism mobilities: Places to play, places in play.* Routledge.

Sheller, M., & Urry, J. (2016). Mobilizing the new mobilities paradigm. *Applied Mobilities, 1*(1), 10–25.

Skoll, G. (2014). *The dialectics of social thoughts.* Palgrave Macmillan.

Stankov, U., & Gretzel, U. (2020). Tourism 4.0 technologies and tourist experiences: A human-centered design perspective. *Information Technology & Tourism, 22*(3), 477–488.

Suliman, S., Farbotko, C., Ransan-Cooper, H., Elizabeth McNamara, K., Thornton, F., McMichael, C., & Kitara, T. (2019). Indigenous (im)mobilities in the anthropocene. *Mobilities, 14*(3), 298–318.

The Island. 2005. Michael Bay. Warner Bros, 136 minutes.

The Matrix. 1999. The Wachowskis. Warner Bros. 136 minutes.

Tzanelli, R. (2015). *Mobility, modernity and the slum: The real and virtual journeys of 'Slumdog Millionaire'.* Routledge.

Tzanelli, R. (2016). *Thanatourism and cinematic representations of risk: Screening the end of tourism.* Routledge.

Tzanelli, R. (2018). *Cinematic tourist mobilities and the plight of development: On atmospheres, affects, and environments.* Routledge.

Tzanelli, R., & Korstanje, M. E. (2019). 6 On killing the 'toured object'. In R. Isaac, E. Cakmak, & R. Butler (Eds.), *Tourism and hospitality in conflict-ridden destinations* (pp. 45–60). Routledge.

Urry, J. (1981). *The anatomy of capitalist societies: The economy, civil society and the state.* Macmillan.

Urry, J. (1988). Cultural change and contemporary holiday-making. *Theory, Culture & Society, 5*(1), 35–55.

Urry, J. (1992). The tourist gaze 'revisited'. *American Behavioral Scientist, 36*(2), 172–186.

Urry, J. (2001). Globalising the tourist gaze. In *Tourism development revisited: Concepts, issues and paradigms* (pp. 150–160). SAGE Publications India Pvt Ltd.

Urry, J. (2002). *The tourist gaze.* Sage.

Urry, J. (2004). The system of automobility. *Theory, Culture & Society, 21*(4–5), 25–39.

Urry, J. (2009). Sociology and climate change. *The Sociological Review, 57*(2_suppl), 84–100.

Urry, J. (2010). Sociology facing climate change. *Sociological Research Online, 15*(3), 145–147.
Urry, J. (2015). Climate change and society. In *Why the social sciences matter* (pp. 45–59). Palgrave Macmillan.
Urry, J. (2016). Does mobilities have a future? In M. Grieco & J. Urry (Eds.), *Mobilities: New perspectives on transport and society* (pp. 3–20). Routledge.
Urry, J., & Larsen, J. (2011). *The tourist gaze 3.0.* Sage.
Vannini, P. (2010). Mobile cultures: From the sociology of transportation to the study of mobilities. *Sociology Compass, 4*(2), 111–121.
Wilkinson, P. (2011). *Terrorism versus democracy: The liberal state response.* Taylor & Francis.

CHAPTER 5

Terrorism and the Allegory of the Lone Wolf: From 9/11 to COVID-19

Introduction

The end of the Cold War and the collapse of the Soviet Union posed new and greater challenges and risks for the Occident; terrorism doubtlessly seems to be one of them. In recent years, and particularly after 9/11, terrorists changed the focus of their attacks. While classic terrorists targeted important persons such as politicians, chief police officers, and celebrities, modern terrorists planned attacks on leisure spots or spaces, tourist destinations, and lay-persons. This is particularly troublesome for policy-makers (who are in charge of orchestrating all-pervading models to preserve homeland security) and for field workers who are seriously punished when they are in contact with radicalized cells. In this context, specialists face many problems in understanding the lone wolves' motivations and goals and their interactions given the latest technological breakthroughs. This chapter interrogates further on terrorism, generally and the lone wolf in particular, and its intersections with colonialism.

The terrorist attacks on September 11, 2001, made significant damage to the US and its economy and shocked the western hemisphere, opening the doors to a securitization process as never before. As a foundational event, 9/11 changed radically the ways nations evaluate their geopolitical policies and the internal checks and balances institutions. Apart from this, scholars are divided into two clear-cut poles while describing the reasons

M. E. Korstanje, B. George, *Mobility and Globalization in the Aftermath of COVID-19*,
https://doi.org/10.1007/978-3-030-78845-2_5

behind 9/11. While some academicians start from the premise that the scourge of terrorism results from contexts of extreme political instability, most of them located in failed states zones or the Third World (Huntington, 2000; Keohane & Zeckhauser, 2003; Mintz & Brule, 2009), others lamented on the systematic ideological manipulation of governments to impose economic policies—if not a neoliberal program—that otherwise would be widely rejected by citizens. To wit, the so-called "War on Terror" centers on a precautionary logic (preemption), which, unless regulated, may affect democratic institutions (Altheide, 2006; Skoll, 2016; Skoll & Korstanje, 2013; Korstanje, 2018a). What is more important, terrorists humiliated the most powerful nation of the world, the US, and directed four commercial airplanes that are considered the hallmarks of Western capitalism against the economic and military core of the West. Over the years, specialists agree with the premise that terrorist attacks were unfeasible for the First World. It is safe to say that 9/11 places this belief in discredit, accelerating a climate of fear and uncertainty worldwide. As Australian sociologist Luke Howie puts it, terrorism inscribes itself in a tensional logic where spectatorship feels captivated but terrorized by the terrorism-related news. Lay-people have been culturally domesticated to consume terrorism as a form of cultural (macabre) spectacle. Far from what social imaginary feels, terrorists do not want a lot of people dying; rather, they want a lot of people watching! Notably influenced by Jean Baudrillard, Howie holds that terrorism should be defined as something else than an act of political violence, but as a semiotic message society receives from a part of citizenship. In this way, terrorists need media to disseminate their message at the time, media packages, and distribute the news to maximize its profits. As a result of this, there is a "phenomenology of terrorism" that transcends cultures and nations worldwide (Howie, 2010, 2012).

Some studies focused on the radical transformation of young second-generation Muslim citizens in recent years. Far from being demons or hatred-filled maniacs, terrorists are lay-people who gradually embrace a radicalization process. Under some conditions, which are studied by psychologists, anyone may be a potential terrorist (McCauley & Moskalenko, 2008, 2017). As Daniele Valentini (2019) observes, terrorists are circumstantially stranded between two worlds: the European and the Muslim ethos. As second-generation Muslims, they face an identity crisis that is finally resolved violently. Having said this, violence catalyzes the tension between a real and a fictional character. Experts who are interested in responding to how a mind is transformed (radicalized) and under what conditions alert that counter-terrorism measures stay one step behind the terrorist cells. Although the specialized literature, in this vein, has advanced in exciting directions, publications can be grouped in two different, but combined, schools: (a) macro-social studies oriented toward explaining the individual behavior through the stick-and-carrot model and (b) those oriented toward understanding the radicalization as a process emotionally based on daily interaction. It is safe to say that radical minds not only are seduced by discourses emanated from marginal groups—in particular because of their negative cosmology to understand the world—but are an easier target to be recruited. The problem with this conception lies in the fact that not all radicals are typically potential terrorists (Dawson, 2009; Borum, 2011; Sedgwick, 2010). As McCauley and Moskalenko (2017) showed, there are 12 clear mechanisms that operate to form a terrorist mind. Radicalization can be successfully achieved when a subject, who developed a negative conception of the external world or society where he or she lives, is brought to live in small groups, geographically isolated from society. In some cases, the recruitment looks to appeal to the emotional basis, shaping a network of peers, relatives, or friends carefully amalgamated into the same cell.

In this backdrop, this chapter critically discusses the archetypical figure of the lone wolf in the Western imaginaries, literature, and cultural forms of entertainment. This continues the previous discussion of the book *The Challenges of Democracy in the War on Terror*, where we stress the methodological limitations of terrorism research today. Since terrorists are considered dangerous criminals, any approach with them is seriously penalized, or, what is worse, ethnographers are pressed by the state to cede their key informant identity. In consequence, the media and the press are filled with "pseudo-experts" who is or are oriented to affirm the status quo. Methodologically speaking, terrorism studies are particularly weak or centered on political (partisan) discourses that, far from solving political violence's philosophical dilemma, support ethnocentric visions (Korstanje, 2018b). The chapter, hence, concentrates efforts on deciphering how the Western imaginary constructs the "Otherness" to explain the formation of lone wolves' minds. In so doing, it combines the postcolonial theory with travel writings derived from the colonial eye, a genre magisterially approached by Marie Louise Pratt, James Clifford, and Marc Augé. Finally, we critically review the novel *La Lenta Agonia de los Peces* [The slow agony of the fishes], originally written by Eric Frattini, to understand how the idea revolving around the lone wolf comes from old colonial stereotypes. To put the same bluntly, terrorism evinces the problems of the West to deal with the "non-Western Other". Lastly, we make some comparison with the recent outbreak of COVID-19 and the closure of several national borders and airspaces, an event that marked a new feudalization of the world.

Let us explain to the readers first that this book is far from being catalogued as a project versed in cultural studies or film studies. However, the media content or film plots and novels serve us to give some fresh and critical insight on issues that otherwise remain out of discussion. Two of the fathers of the Frankfurt School, Max Horkheimer and Theodor Adorno, elicit an interesting critique on the ideals of Enlightenment. Their book, entitled *Dialectic of Enlightenment*, saw the light of publicity in 1947. The mainstream cultural values of the Enlightenment have been forged in the colonial period and of course have perpetuated to capitalist society through rationalism and the liberal society. Over centuries, the native was seen as a passive source of reception of the illustrated European thinking accepting passively a message they were unable to reverse. Adorno and Horkheimer are moved to expand the understanding of Nazi Germany's socio-economic background and the devastating impacts on European societies just after the end of World War II. At a closer look, the book combines part of Adorno's works, in particular his studies on authoritarian personality, with a caustic Horkheimer's philosophy. The classic dominant paradigms in social science explain neither the authoritarian personality nor the radicalized nationalisms. The rise of Hitler to the presidency seems to be because of a much deeper socio-economic background which was finely enrooted in Europe. Of course, the ideals of Enlightenment reflect part of this radicalization against the "non-Western Other". These nationalisms are based on irrational superstitions which reinvented mythology. In the Enlightenment, the message is unilaterally fabricated by the scholarship and accepted by the lay-persons. There is only one truth, which is inexpugnable to the human ordinary eye. During the process of Enlightenment, the critical philosophy was over-rationalized to become an instrument of the market. Likewise, the modern man is trapped between a rationalized means of production and the culture industry. In the same way, mass culture standardizes, if not commoditizes, daily life in cultural goods since the standardized products are disposed to domesticate citizens. If the telephone allowed a certain reply, the mass medium that started with radio and TV closes the connection of the audience with their interlocutors. Invariably this leads to passive listeners who passively receive the message. Hence scholars should penetrate the ideology of capitalism through critical thinking (Adorno & Horkheimer, 1997). In a world transformed in a phantasy, phantasy offers a fertile ground to understand reality.

A Return of the Non-Western Other and Colonialism

Without any doubt, James Clifford is an authoritative voice in colonization theory. He has shed light on the intersection of geography, travels, and the imperial European powers. Per his viewpoint, journeys reflect a combination of two opposed cultural values: dwelling (here) and moving forward (there). In contrast to other scholars, Clifford starts from the premise that travels and knowledge are inevitably entwined. The term *theory* stems from the Greek *theorein*, which means the practice of observation, a ritual where a free man was dispensed to travel to another city to gaze at a religious event. This ritual of validation denotes the real nature of theory as a residual consequence of a physical (and today intellectual) displacement. The colonial cosmologies, which molded a binomial tension between here-there, us-them, or West-East, seem to be nowadays in constant change. The quest for something novel plays a leading role in the configuration of historical reasoning, which is associated with writing. We need to travel not only to test our theoretical assumptions but to document what we gaze. This is the common point between history and travels, as Clifford contends, which has been radically changed by the expansion of globalization. The notion of alterity and the idea of out there has been mutated (Clifford, 1989, 1997a, 1997b).

As previously argued, travels and journeys have occupied a central position in Western history. From Julius Caesar to Max Weber, theorists have visited strange lands and geographies to understand their societies and the "Other" as a mirror of their own culture. Nonetheless, such a meeting was characterized by hard conditions where the European travelers were not always welcomed. To some extent, for the European social imaginary, technology embodies the archetype of civilization creating a double hermeneutic between the conqueror and the native (conquered). To put it differently, technology facilitated the geographical displacements of military forces to pacify the peripheries; writers, ethnographers, scientists, and business travelers paradoxically affirmed the so-called superiority of European mainstream cultural values over the immobile natives. The authority of European writers emanated from the inferiority natives passively accepted (Rojek, 1993; Grewal, 1996). To explain better the issue, as Marvin Harris brilliantly states, one must pay attention to two (combined) factors. On the one hand, the rise of the anthropological theory coincides with the expansion of political and economic needs in Europe (from the sixteenth and seventeenth centuries onward). At least as it was originally imagined, the anthropological theory recreated the conditions for a dialogue with the "non-western Other", when it gave vital information to colonial governors. Colonial officers devoured texts and books written by ethnographers to get basic details on warfare tactics, customs, and aborigines' strategies. Aligned in the same campaign though, for different reasons, anthropology and colonialism were inextricably intertwined. On the other hand, European travelers (scientists) were educated in a climate of paternalism where the alterity was culturally seen as inferior but with good (promising) possibilities to be re-socialized according to the highest European ideals. The picturesque figure of the noble savage offers a romantic and idealized image of the native, which does not break with the idea that he is a wild man but, what is more important, provides the essence toward a romantic primitivism, widely inspired by cultural genocide.

The first ethnographers strongly believed that, like in pre-modern Europe, native cultures in the overseas territories were doomed of final disappearance. In early Europe, countless feudal structures disappeared before the advance of modernism and industrialism. The notion of modernity was a vehicle toward the economic process but, paradoxically, to moral corruption and greed. Different social maladies, such as crime, delinquency, anomie, or alcoholism, emerged as a direct consequence of modernity. The archetype of the noble savage epitomizes the deep interplay between the pristine, or pure, moral man, who ethnographers should protect, and the cruelty of conquerors or colonial authorities. Hence, anthropologists were certainly worried and pressed to collect as many artefacts, stories, and relics as they can. No less true is that the paradox lies in the fact that the urgency for protecting the "non-Western Other" finally legitimated the colonial conquest (Harris, 2001).

As already discussed, *Imperial Eyes: Travel Writing and Transculturation*, Pratt's book (2007), gives an interesting insight that describes how beyond the needs of knowledge lies the logic of what she dubbed as *Imperial Eyes.* Since European paternalism would never be feasible without the available technology that disposed travels and expeditions, the interest in science was a critical factor that determined the successive journeys. Europeans moved not only by curiosity or the quest for novelty but by the needs to impose a classificatory system that describes the sensible world. In this context, travel writing ignited a dormant romanticism that played a leading role in the formation of European ethnocentrism. Some texts denounced the injustices and cruelty of the colonial order, while others the lack of decency of natives (Pratt, 2007). As Pratt remarked, the imperial eyes sailed in a double dimension: while they incarnated a dominant position through the authority of scientific knowledge, natives internalized so-called cultural inferiority regarding European reasoning, packaging a unilateral view of history. In this instance, Pratt helps to unpack the much deeper connection between modern science and colonial authority.

The quest for novelty, which so to speak inspired the European travelers, not only operated in the construction of a biased history but contributed to the configuration of the modern nation-state, at least as we know it today. In a nutshell, she gives interesting analytical instruments to understand the alterity's role as a sub-product of travels and geopolitical cosmologies that emanated from European logic (Korstanje, 2018b). As Nichole Guidotti-Hernández (2011) puts it, the nation-state, as a hegemonic project, imposed through the articulation of *unspeakable violence* which ideologically determined sentiment of belonging over nomad groups. This violence was associated with nation-state and nationality archetypes, subordinating the local native to a specific territorial organization. Like history, the law marked the newly emerging political authorities as well as the material asymmetries between the European lords and natives. In this vein, as Guidotti-Hernandez explains, the evolution of nation-states corresponds with a territorial organization, carefully orchestrated in an ethnic hierarchy, where differentiation was successfully imposed through racism and classism. The idea of the "Other" gave coherence to the local agents homogenizing the national character so that states exert control over their citizens. Citizenship was shaped by selecting historical facts (events) rememorized once again through national symbols.

The European matrix constituted a new cosmology to understand the world, where the "non-Western Other" should be protected but subordinated to the colonial rationality. Just after the end of World War II, peripheral nations faced a set of riots, civil wars, and political instability processes, which invariably led them to the need for separation and independence respecting the colonial metropolises. The emerging social discontent echoed from the urgency of adopting democracy as the main form of government ground colonialism to a halt. While the old center-periphery dependency now set the pace for a new order, the Global North founded the background toward a new domination instrument: development theory, which suggested that the industrialized nations needed to bail out poor unindustrialized economies. Just after Truman's speech, the world was split into two poles: developed and underdeveloped countries. The archetype of the noble savage, which in former centuries oiled the imperial machine, was now reframed in the shape of economic progress. Paradoxically, at the time European nations began to colonize the overseas territories, thousands of migrants abandoned their home to take a better chance in the first world, forming a new powerful workforce that constituted the urban social scaffolding (Korstanje, 2018a, 2018b). Marc Augé has observed that global modernity forced ethnographers to rethink the borders of here-there or selfhood and otherness because of an identity crisis. Centered on his fieldworks in Africa, Augé holds a more than interesting—but not for this less polemic—argumentation. Since a place can be, at least anthropologically speaking, a site of common belonging, tradition, or history, where the subject is conditioned to its own constituted biography, *non-places* are mainly characterized by a process of social destruction, indifference, and depersonalization, where the biographical subjectivity is eradicated. In the non-places, while mass consumerism and depersonalization are not the exceptions but the rule, identity is electronically verified in specific check-in spots. Examples of non-places are airports, transport hubs, shopping malls, to name a few. The quest for consuming heritage, probably what tourists are obsessed with, is an implicit way of eradicating history (Augé, 1995a). The cosmopolitan spirit, which is characterized by the taste of high culture, makes social relationship vague, native, and unsubstantiated. At its outset, anthropology was seriously worried by the vulnerability of natives, but this has finally changed because peripheral natives opted to migrate to urban cities. Globalization altered the geographies of nation-states, laying the foundation toward the creation of a new (flat) territory where the neighbor becomes a stranger for us. The sense of familiarity has set the pace for a global landscape where here and there have been blurred.

The modern tourism industry has paved the way for digital technologies to accelerate transport timeframes and travel costs. As a result of this, the Occident expanded rapidly to the four corners of the planet. The earth was flattened while cultures were homogenized in the name of consumption. No matter what the ethnicity or the culture, all humans are subject to the same standardized lifestyles which are subordinated to the market rules. The disappearance of nations made citizens become strangers to themselves (Augé, 1995b). Although Augé does not focus on the archetype of the lone wolf, his commentaries very well apply to the problem of Muslim migration and terrorism. Plots of movies or even novels portray jihadists as resentful citizens who embrace the Muslim faith because they hate society. Before being trapped between two opposing worlds, terrorists had a brilliant future; they are even helped and supported by the state in their education, but once grown, they deal with a radical rupture with normal life, certainly forced by a traumatic event or moved to experience a new revelation. Terrorists, having said this, move toward a spiritual trip to the Middle East to have a renovation. Once met, the experience, which is determined by the foundational (ritualistic) trip, changes their minds. Not surprisingly, they come back but not in the same way; they are now "jihadists" or "warriors of the lord" fighting against the cruelty and corruption of the West. After disappearing like anvils under the waves, they emulate the sinew of war in their hearts. The process of radicalization has started.

Narratives like this occupy a central position in the Western social imaginary and legitimize the rise of violence against Muslims as well as Islamophobia. The sentiment of terror instilled by terrorism is ideologically manipulated to trace, monitor, and trial certain ethnic minorities while violating their basic rights and the powers of checks and balances that warrant the democratic life (Korstanje, 2018b). As Ritwik Balo and Sapataparni Sadhu (2019) discuss, the exegetes of state are educated to believe the state reserves the right of violence when political stability is in danger, but the conditions from where this violence is or not exerted are not well defined. The violence of the state seems to be always legal, though in some cases it violates human integrity. As Balo and Sadhu said, Adolf Hitler committed horrendous but legal crimes. This begs an interesting question: who is the judge when the law is finally adapted to the rules of victors? The reasons why literature helps scholars calibrate a critical focus on the dilemmas of terrorism are based on the power of imagination and fictionality as valid interlocutors that interrogate reality's injustices. Finally, Dashielle Moore (2019) calls attention to literature as the only way of penetrating, if not resolving, national banality's philosophical problems. The ultra-patriotic movements impose some banal thinking that feeds up the social media and the cosmologies of the post-9/11 context. The triumph of popular visual culture—without mentioning the cultural entertainment industry—to gain firm steps toward the representation of terrorism has pushed many novel writers and poets to a marginal position. Those writers opposed to the dominant discourse were accused of being national traitors or demonized as sympathizers with jihadists.

The Lone Wolf

Some voices alert that terrorism studies go through a state of stagnation, which leads gradually to an epistemological crisis (Sageman, 2014). In a landmark book entitled *Formations of Violence*, Allen Feldman (1991) dissects the hermeneutics of terrorism, focusing on the tactics of the Irish Republican Army and the counter-terrorism measures headed by the British Army in Northern Ireland. For Feldman, terrorism should be understood as a political process aimed at depersonalizing citizens. In normal conditions, the figure of a traditional Irishman fighter speaks to us of a person who exerts violence with his body, but the terrorist commits a crime remotely, only using an external device. In the attack, the terrorist dissociates his body from his objectives. For that reason, as Feldman adheres, terrorism evinces a type of social decomposition, noting a decline of social ties.

As Lisa Stampnitzky (2013) puts it, terrorism research has faced a stage of significant fragmentation, where there is no consensus to define what terrorism means, and once the discipline evolved toward a facet of maturation, the phenomenon was morally condemned. In the 1970s, terrorism was associated with insurgency or the clash between two asymmetrical armies. Today, terrorists are seen as demons, radical fundamentalists who hate the Occident and its lifestyle. Above all, this represents a serious methodological problem for ethnographers who are in contact or have key information from terrorist cells. Luke Howie (2012) infers, they are often pressed to make public the identity of key informants who are considered dangerous enemies of society. The gap left by the demonization of terrorists and the urgency to understand the issue is normally filled by "pseudo-experts" who enthusiastically embrace the precautionary logic aimed at ghettoizing some ethnic minorities. The process seems to be accompanied by a securitization campaign where the "non-Western Other" is potentially suspected to be a terrorist. In this way, the government often evokes the principle of security to impose neoliberal policies; otherwise, it would be amply rejected (Altheide, 2004, 2007; Jackson, 2009; Skoll, 2016;

Korstanje, 2018a). To some extent, the executive branch accumulates a centralized power while passing over other branches and checks and balances institutions. The so-called War on Terror continues the tradition of identifying an internal enemy to close ranks behind a strong leader. The war on terror is not pretty different from the war on cancer or the war on crime (Simon, 2007). Beyond the enigma of the lone wolf, a great number of books and publications were published, but far from being scientific works, they affirmed the circulating stereotypes of social imaginary (Spaaij, 2010; Moskalenko & McCauley, 2011). Enders and Sandler (2011) overtly say that terrorism is motivated by economic interests. As rational agents, terrorists look to maximize their gains while minimizing the inherent risks. Today's terrorists target leisure spots or tourist resorts because of their psychological impacts. Since lay-persons populate these spaces, the social imaginary believes that anyone and anytime would be a potential victim of terrorism. It instills a climate of terror that presses the government to accept the terrorist claims. At a lower cost, terrorist receives higher gains. In a seminal book, *The Enigma of Lone-Wolf Terrorism*, Ramon Spaaij (2010) provides a clear diagnosis that explains the radicalization process's socio-cultural factors. By the same token, governments devote efforts and financial resources to curb the activity of radicalized cells that often recruit candidates who ultimately become lone wolf terrorists. These candidates even are recruited through digital media and social networks. What is equally important, digital technologies offer lone wolf terrorists an essential part of interacting with peers and terrorist groups. Meanwhile, they pretend to lead a normal life. Based on a compilation of lone wolf terrorism in 15 countries, from 1968 to 2010, Spaaij argues convincingly that the outstanding commonality of these terrorists seems to be their normality, which means they lack any previous psychopathology, as the media emphasizes. Nonetheless, there is a strong and deep process of radicalization, which is culturally determined. In this vein, Peter Phillips (2011) acknowledges that the current research, or at the least the state of the art, exhibits no few methodological problems to explain how a radicalized mind is shaped. In this backdrop, neither economic analysis nor government security policies suffice to describe the lone wolf acts.

At a closer look, lone wolves are physically isolated or retired temporarily from their daily activities, exiled abroad over the years. Their vulnerability is often given by the socio-economic status and the countless deprivations they are immersed in. Policy-makers should remember that these potential terrorists can be re-educated to take part in the community. Ian Baptiste (2001) gives an interesting model to re-socialize young people who have taken part in terrorist attacks. Per his viewpoint, the meritocratic ideals, which are characteristic of capitalism, socialize citizens to be more efficient or more productive. The meritocracy recreates the conditions to solipsism, individualism, and greed, leading to material asymmetries where terrorism becomes stronger. However, James Piazza (2006, 2008), in different works, contradicts such an axiom, arguing that neither democracy nor poverty is the key factor that predisposes the subject to terrorist acts. As he acknowledges, though poverty and frustration may be very well sublimated against an external enemy, terrorists accept a type of pseudo-historicism based on unreal (fabricated) events. In the terrorist mind, historical facts are not remembered as they happened but in a biased way. The radicalization works simply because the goals of the terrorist group are vague and almost impossible to achieve. Suffice to say that jihadists look for impossible ends, which oscillate from the end of the West or capitalism to a global Caliphate. In an interesting article, Moskalenko and McCauley (2011) enumerate a set of variables that correlate directly with radicalization. Per the obtained outcome, they hold that lack of empathy with victims associated with much deeper victimization because imagined injustices play a leading role in the radicalization of jihadists. Lastly, some recent studies focused on the role of self-esteem and in-group rules as determinant factors of terrorist recruitment. Young terrorists are often selected or seduced by friends, peers, or relatives who take part in the organization. It is safe to say the emotional factor occupies a central position in the recruitment process (Knapton, 2014; Coppock, 2014; Kruglanski et al., 2013; Adekeye et al., 2018). All these above-discussed models give a snapshot of the problem, but they lack a robust basis to expand a holistic understanding of this much deeper issue. In filling this gap, we, in the next section, offer a qualitative view of the novel *La Lenta agonia de los peces* [The slow agony of the fish], which proffers a firm basis to explain how the allegory revolving around the lone wolf is culturally drawn. Having said this, literature plays a crucial role in deciphering the dominant discourses.

La lenta agonia de los peces

There is a strange paternalist discourse in the Western social imaginary aimed at presenting capitalism as the best of feasible worlds. The non-Western Other is ambiguously welcomed but frightened as a potential danger. This moot point is debated in this section through the novel authored by Eric Frattini. As noted, Frattini does not need an introduction. He has vast experience as a journalist as well as a war correspondent in the Middle East. His novels and texts have been translated into more than a dozen languages, and in recent years, they have focused on the efforts of Western governments to defeat terrorism. This is his main concern, found throughout his novel *La Lenta Agonia de Los Peces*, which ESPASA published in 2013. The plot centers on the life of the fictional character Havana Sinclair, who is a brilliant counter-terrorist agent at the services of British intelligence. Even if she is one of the most prominent experts in Islamic terrorism who always sees matters through, she is fired from her unit, MI5, after the 7/7 London bombings that occurred in 2005. Havana meets a private insurance company that requests her experience and services in Baghdad. She is asked to investigate the murders of two millionaire investors. Once there, she struggles against old foes. The story situates around a real-life terrorist attack in London, England, where four bombings killed 52 civilians. The first explosion occurred at 8.50 a.m. between Liverpool Street Station and Aldgate, while two other bombs were detonated at Edgware Road and Russell Square. The last attack was on a double-decker bus in Tavistock Square. One of the terrorists was Hasib Hussain, a British Muslim whose parents came from Pakistan. His father served as a rank-and-file factory worker, and his mother aided newcomers and migrants with their English. Not only was Hasib lucky enough to go to high school, but he was also known as a smart boy, very polite, and gifted. Here the question surfaces as to what happened to make Hasib become a terrorist.

Whatever the answer may be, the plot indicates that he was voluntarily involved at a mosque where he met some radical terrorists. He is suspected to have visited Pakistan in recent years, but nobody knows anything about Hasib anymore. In parallel, Havana is fired from her MI5 position. What is more interesting in Frattini's argument is that London was selected as the host city for the 2012 Olympic Games the day before the bombings. These four strikes left 52 victims and more than 700 were wounded. Such a traumatic event cost Havana Sinclair her position as an agent and terrorism expert. As an incorruptible woman, Havana travels to Baghdad to look into the above-mentioned crimes. She moves in the shadows of British intelligence, questioning herself to what extent she takes part in the correct band. Frattini, who served as a well-known expert in terrorism, imaged fiction which very well has been compared to reality.

The novel gradually ushers readers into the world of Islamic terrorism, alternating some aspects of fiction with the current challenges Europe faces today. A deeper mistrust of government accompanies the fear of terrorism. Frattini's novel is centered on an ethnocentric discourse that depicts the Muslim community, or, to be more precise, Muslim Britons, as potential terrorists. The novel ideologically portrays Hasib as a man who has everything his parents lacked. He is smart, born and well educated in a first-world country, but this does not suffice for him. In quest of a more authentic lifestyle, he needs to get closer to religion to develop England's intricate hatred. This raises an interesting question, is religion the key factor in Muslim terrorism?

In contrast with Hasib, Havana is a mere British citizen who is willing to sacrifice her life for her nation and fights against dark powers to protect homeland security without asking anything in return. She is blamed for what happened but she still does what should be done. Once in Iraq, where the American army has deployed thousands of soldiers, she meets Jalid, who will be her driver and escort. This city is dangerous for everyone, and particularly for women. A commando group finally kidnaps Havana in Mutanabbi borough. Frattini (2013) portrays a grim landscape of Iraq, a country devastated by war.

> Havana, todavía con la capucha puesta sobre la cabeza, olía los medados y excrementos que se amontonaban a su lado. Las moscas se posaban sobre su piel y su ropa, pagada al cuerpo por el sudor, ante la atenta mirada de un adolescente que la vigilaba con un rifle sobre sus piernas. (p. 153)
>
> [Havana, hooded on her head, smelt the piss and shit accumulated along her, while flies landed on her skin and clothing, which was stuck to her body by the extreme sweat. Before her, a teenager kept a gun between his legs.]

Like many other teenagers in Iraq, for Frattini, this teenager lives in a devastated country; doubtless, he is a victim but at the same time a killer who assassinates following the orders of maniacal leaders. Iraq is a site of war, and devastation, where children have no future. The earlier regime executed his father, and his two brothers were unjustly killed by the current government. As an insurgent, the boy knows one of his brothers was jailed at Abu Ghraib while the other lost his life as a suicide bomber in Zona Verde's Wall (Green Zone). While in captivity, Havana realizes how human rights ideals, as they are imagined in the West, are systematically violated in the periphery. In this respect, Havana is hosted with a contractor, who comments that they will all die. She answers that they have no reason to think this, but the trembling businessman tells her: "soy contratista, sabes que significa esto en Irak? Desde que nos capturaron a Charlie y a mi, me han humillado y me han torturado" (p. 174). [I am a contractor, do you know what that means in Iraq? Since Charlie and I have been kidnapped by these guys we have been humiliated and tortured.]

Though Havana is confident that she will be liberated, the contractor tells her during their long conversations that he is a hitman contracted to target some specific persons. This secret surprises Havana, who suspects that being there is not a concidence to just be there after all. He was commanded to assassinate two hunters, Tibbals and Mcmillan, who loved hunting human beings as trophies for sport. Havana is unsure whether she will live, but what is most important, she sees the grim future of Iraqis, creating a sentiment that leads her to confront her Western presumptions. She is finally released by an unknown person and found on the street by a British patrol. She misses her husband (Peter Sinclair), who represents the British oligarchy, and her daughter. She is faced with serious dangers and risks on her return to England. After all, those who had carefully planned the attacks in London were monitoring her steps. Havana not only discovers a dark conspiracy planned by high-ranking British intelligence officials such as Raymond Clarke, but she realizes the plans were orchestrated to change the internal politics of the UK. In his novel, Frattini depicts a typical landscape. Some corrupted agents located at the top of the pyramid of the security forces or the government who should care for citizens handle Muslim terrorists to encourage a new social order.

To cut a long story short, *La Lenta Agonia de los Peces* presents as a well-written novel whose main theme is terrorism. At a first glance, Frattini follows an ethnocentric discourse that has the following elements: a strong and incorruptible hero (Havana) who sacrifices her life and comfort, a constituent of the First World, to protect others. This sacrifice takes her to remote and lawless nations where human rights are violated daily and where people die for no reason. The pre-Hobbesian state of chaos and violence that characterizes Iraq, which is a zone occupied by the Anglo-alliance, seems to be one of the grim landscapes Frattini often draws. On the other hand, Havana's husband emulates the elite's passivity, which declares itself unable to fight against terrorism. The perpetrators of the London bombings were first-generation Britons whose parents were welcomed under the sacred law of hospitality. Hasib appears to be a brilliant and promising student who embraced a sentiment of hate and radicalization against the UK, risking in this way his opportunity to have a better life. Here we find some of Howie's comments respecting *the unsaid terror*, which is shaped by the articulation of countless complex narratives. Terrorists live like us, they are infiltrated in the core of Western nations, they may even be our neighbors or, what is worse, friends. This puts all of us in a difficult position because anyone may be a victim of terrorism, anytime and anywhere.

Is COVID-19 a Foundational Event?

The 9/11 and COVID-19 particularly share commonalities and serious differences. At first glance, the numbers 1 and 9 repeat mysteriously in both; however, these are mere, if not naïve, speculations. The original effects of terrorism in the US were the stagnation of trade and the cancellation of all flights and travels. This generated serious losses for the US. Even George Bush called attention to the need for Americans to travel abroad and not to succumb to the terror instilled by Al-Qaeda. This foundational event marked a serious crisis for Western hospitality simply because it was the first blow planned using four commercial airplanes: the bulwark (the signs) of American capitalism. In the successive months, of course, migration bans and laws of homeland protection were sanctioned followed by a strict method of control at borders and airports and migratory visa requirements. The "Other" gradually became an object of fear and suspicion. American tourists avoided travelling abroad, in particular to unknown destinations, preferring to travel within the US. In consequence, domestic tourism prevailed during a couple of months over other forms of international travel. Here is the first difference between COVID-19 and 9/11. The pandemic forced to shut down all airports worldwide, declaring cancellation of all commercial flights and the closure of borders and airspace. The tourism industry ground to an unparalleled halt. Governments began to issue loans and financial credits to bail out corporations from inevitable bankruptcy. What is more important, there was no foreign country to blame or to attack, like Afghanistan or Iraq. Probably China occupied such a position and the WHO because of the irregularities in the communication process during the pandemic.

Another difference is associated with the nature of this global risk. While terrorism speaks often of an organized cell that operates in clandestine life, the virus seems not to be alive. Having said this, anyway, both events share the same consequence: the decline of Western hospitality.

COVID-19 prompted long-dormant geopolitical crises among states (e.g., between the Netherlands and Spain) because of the financial budgets and old forms of racism and chauvinist discourse, above all Chinese tourists emerged. Asian tourists were originally seen as a source of infection, mistreated, rejected, and sometimes roughly treated. The same occurred in China with the Western tourists, once the first wave was successfully placated. In the days of 9/11, anyone could be a potential terrorist, even a friend or a neighbor. Now, we all are real terrorists (carriers) of a lethal disease who can very well kill other citizens. Tourists and citizens are suddenly forced to live in isolation, making the home a jail. People need to wear a mask and cover their mouth to prevent the COVID-19 infection and of course to stop the mass contagion. Nobody knows who has been infected, in which case the precautionary logic dominates the day-to-day politics. Without knowing (asymptomatic or in a mild form of the disease) anyone can be infecting others. The first contact is always done through the eyes. In the post-pandemic context, we are unable to identify visually the "Other".

In earlier chapters, we used the term *wicked gaze* to explain locals' lack of empathy with tourists. This was not new. Not surprisingly, anti-tourist sentiments flourished in urban tourist cities as Barcelona, Venice, and Rome, but in this case, tourists were overtly discriminated against or expatriated without qualms. The nations not only closed their borders but entered into discrepancies internally to keep the internal frontiers. Local governors made the unconstitutional decision to close domestic borders, impeding free circulation; in some cases some governors adopted a separatist attitude respecting the central administration. What is more important, the social contact that characterizes human nature without mentioning the celebration's anthropological needs was temporarily suspended. Feasts, events, marriages, and any other form of celebration were strictly prohibited. It is safe to admit that 9/11 and COVID-19 share another commonality; the anxieties amounted to an invisible enemy who cannot be grasped through the gaze, an enemy who looms everywhere and can attack anytime. In this vein, we refer to an invisible foe, who confronts the discourse of ocularcentrism, which molded the West civilization's spirit. Hospitality started to die just after 9/11, weakening or disorganizing the social ties, creating a discourse based on fear while closing the opportunities to meet with the "Other". COVID-19, far from reversing such a tendency, paves the way for the rise of new normality plagued by domestic disputes, territorial fracture, geopolitical tensions, and the elaboration of narratives oriented toward demonizing foreigner tourists. The following chapters will provide further well-grounded details of this assumption for further clarity.

Conclusion

This chapter debates the decomposition of the social fabric, which paved the way for the rise of conspiracy plots and the new culture of the entertainment industry that packages terrorism as a form of spectacle. Through the articulation of different ideological narratives, the ruling elite imposes a biased image of the Muslim community as a Trojan Horse. The figure of the non-Western "Other" partially accepted in the core of Western civilizations embodies the Muslim whom the West daily neglects. As a continuation of the colonial order, the liberal state, from its outset, is embraced to discipline and inoculate the external dangers—paraphrasing Foucault—in the same way a vaccine is obtained after dispossessing the virus of all its destructive features. Nonetheless, the turn of the century adjoined to some long-dormant fears (of aliens), adjudicated to 9/11 and the so-called compliance of Muslims with terrorism a mantle of suspicion. Not fully integrated into the local community, unlike in Latin American countries, since the Syrian migration fluxes dated back to the nineteenth century, in the Global North, Islam is an object of fear and anxiety. Echoing the medical gaze metaphor, which is designed to locate and eradicate the disease, Western rationality does not understand the grey zones. We used the example of cancer, which is potentially seen as the degeneration of cells.

Western rationality will look at saving the body, extirpating the affected organ if necessary. This appears to be the main reason why the Muslim community runs a serious risk unless significant plans of contingency are orchestrated against terrorism. This chapter was mainly oriented toward discussing the socio-economic background of terrorism and its effects, unearthing old lessons Latin Americans remember. Terrorism centered on the logic of instrumentalization, which is masked in the heart of Western rationale. Democracy, in this case, takes different shapes according to the invocations of the political parties. This chapter described and discussed the different theories around democracy and explained how terrorism might place it in jeopardy. We reviewed the origin and evolution of liberal doctrine, which was functional to the creation of a liberal state to understand the world of consumption. The allegory of *lone wolf terrorism* is functional to these Western-generated discourses which date back to the colonial rule. However, 9/11 introduced radical shifts to the industrial world. The enemy, now, does not reside there in the wild (unknown) colony; rather they live with us. This suggests that hospitality as the foundational (sacred) value of the West has been undermined.

Meanwhile, we have laid the foundations for a new understanding of the fear of the stranger and the psychological effects of the recent COVID-19 virus outbreaks in the social imaginary of the West. We hold the polemic thesis that far from being seen as a *foundational event*, the pandemic affirms a tendency that originated in 2001, after the attacks on the World Trade Center and Pentagon. This event, doubtless, marked the start of a new world where the "Other" gradually becomes an *undesired guest*. The point of entry is accompanied by what we dubbed as *the death of hospitality*.

References

Adekeye, O. A., Igbokwe, D. O., Agoha, B. E., Olowokere, E. I., Adeusi, S. O., & Elegbeleye, A. (2018). Terrorism inclination and self-esteem level of secondary school students in Nigeria. *Covenant University International Journal of Psychology, 3*(1), 21–32.

Adorno, T. W., & Horkheimer, M. (1997). *Dialectic of enlightenment* (Vol. 15). Verso.

Altheide, D. L. (2004). Consuming terrorism. *Symbolic Interaction, 27*(3), 289–308.

Altheide, D. L. (2006). Terrorism and the politics of fear. *Cultural Studies, Critical Methodologies, 6*(4), 415–439.

Altheide, D. L. (2007). The mass media and terrorism. *Discourse & Communication, 1*(3), 287–308.

Augé, M. (1995a). *Non-places: Towards an anthropology of supermodernity* (J. Howe, Trans.). Verso.

Augé, M. (1995b). *Hacia una antropología de los mundos contemporáneos.* Gedisa.

Balo, R., & Sadhu, S. (2019). 'But you live in Kashmir': Is one person's terrorist another's freedom fighter? In B. Majoul (Ed.), *Terrorism in literature* (pp. 56–71). Cambridge Scholar Publishing.

Baptiste, I. (2001). Educating lone wolves: Pedagogical implications of human capital theory. *Adult Education Quarterly, 51*(3), 184–201.

Borum, R. (2011). Radicalization into violent extremism I: A review of social science theories. *Journal of Strategic Security, 4*(4), 7–36.

Clifford, J. (1989). Notes on travel and theory. *Inscriptions, 5*(5.29), 11.

Clifford, J. (1997a). *Routes: Travel and translation in the late twentieth century.* Harvard University Press.
Clifford, J. (1997b). Spatial practices: Fieldwork, travel, and the disciplining of anthropology. In *Anthropological locations: Boundaries and grounds of a field science* (pp. 185–222). University of California Press.
Coppock, V. (2014). 'Can you spot a terrorist in your classroom?' Problematising the recruitment of schools to the 'War on Terror' in the United Kingdom. *Global Studies of Childhood, 4*(2), 115–125.
Dawson, L. L. (2009). The study of new religious movements and the radicalization of home-grown terrorists: Opening a dialogue. *Terrorism and Political Violence, 22*(1), 1–21.
Enders, W., & Sandler, T. (2011). *The political economy of terrorism.* Cambridge University Press.
Feldman, A. (1991). *Formations of violence: The narrative of the body and political terror in Northern Ireland.* University of Chicago Press.
Frattini E. (2013). *La lenta agonia de los peces - The slow death of the fishes.* Barcelona, Espasa Ed.
Grewal, I. (1996). *Home and harem: Nation, gender, empire and the cultures of travel.* Duke University Press.
Guidotti-Hernández, N. M. (2011). Unspeakable violence. In *Remapping US and Mexican national imaginaries.* Duke University Press.
Harris, M. (2001). *The rise of anthropological theory: A history of theories of culture.* AltaMira Press.
Howie, L. (2010). *Terror on the screen: Witnesses and the re-animation of 9/11 as image-event, popular culture.* New Academia Publishing, LLC.
Howie, L. (2012). Witnessing terrorism. In *Witnesses to terror* (pp. 155–175). Palgrave Macmillan.
Huntington, S. P. (2000). The clash of civilizations?. In *Culture and politics* (pp. 99–118). Palgrave Macmillan.
Jackson, R. (2009). The study of terrorism after 11 September 2001: Problems, challenges and future developments. *Political Studies Review, 7*(2), 171–184.
Keohane, N. O., & Zeckhauser, R. J. (2003). The ecology of terror defense. In *The risks of terrorism* (pp. 103–131). Springer .
Knapton, H. (2014). The recruitment and radicalisation of western citizens: Does ostracism have a role in homegrown terrorism? *Journal of European Psychology Students, 5*(1), 38–48.
Korstanje, M. E. (2018a). *The challenges of democracy in the war on terror: The liberal state before the advance of terrorism.* Routledge.
Korstanje, M. E. (2018b). *The mobilities paradox: A critical analysis.* Edward Elgar Publishing.

Kruglanski, A. W., Bélanger, J. J., Gelfand, M., Gunaratna, R., Hettiarachchi, M., Reinares, F., ... Sharvit, K. (2013). Terrorism—A (self) love story: Redirecting the significance quest can end violence. *American Psychologist, 68*(7), 559.

McCauley, C., & Moskalenko, S. (2008). Mechanisms of political radicalization: Pathways toward terrorism. *Terrorism and Political Violence, 20*(3), 415–433.

McCauley, C., & Moskalenko, S. (2017). Understanding political radicalization: The two-pyramids model. *American Psychologist, 72*(3), 205.

Mintz, A., & Brule, D. (2009). Methodological issues in studying suicide terrorism. *Political Psychology, 30*(3), 365–371.

Moore, D. (2019). What happens in Small rooms: deLillo and the terrorist spectre. In B. Majoul (Ed.), *Terrorism in literature* (pp. 96–112). Cambridge Scholar Publishing.

Moskalenko, S., & McCauley, C. (2011). The psychology of lone-wolf terrorism. *Counselling Psychology Quarterly, 24*(2), 115–126.

Phillips, P. J. (2011). Lone wolf terrorism. *Peace Economics, Peace Science and Public Policy, 17*(1), 1–28.

Piazza, J. A. (2006). Rooted in poverty?: Terrorism, poor economic development, and social cleavages. *Terrorism and Political Violence, 18*(1), 159–177.

Piazza, J. A. (2008). Do democracy and free markets protect us from terrorism? *International Politics, 45*(1), 72–91.

Pratt, M. L. (2007). *Imperial eyes: Travel writing and transculturation*. Routledge.

Rojek, C. (1993). *Ways of escape: Modern transformations in leisure and travel*. Springer.

Sageman, M. (2014). The stagnation in terrorism research. *Terrorism and Political Violence, 26*(4), 565–580.

Sedgwick, M. (2010). The concept of radicalization as a source of confusion. *Terrorism and political violence, 22*(4), 479–494.

Simon, J. (2007). *Governing through crime: How the war on crime transformed American democracy and created a culture of fear*. Oxford University Press.

Skoll, G. R. (2016). *Globalization of American fear culture: The empire in the twenty-first century*. Palgrave Macmillan.

Skoll, G. R., & Korstanje, M. E. (2013). Constructing an American fear culture from red scares to terrorism. *International Journal of Human Rights and Constitutional Studies, 1*(4), 341–364.

Spaaij, R. (2010). The enigma of lone wolf terrorism: An assessment. *Studies in Conflict & Terrorism, 33*(9), 854–870.

Stampnitzky, L. (2013). *Disciplining terror: How experts invented 'terrorism'*. Cambridge University Press.

Valentini, D. (2019). Tabish Khair's just another Jihadi Jane: Inside the mind of a foreign fighter. In B. Majoul (Ed.), *Terrorism in literature* (pp. 20–43). Cambridge Scholar Publishing.

CHAPTER 6

Mobilities and Globalization in the Aftermath of COVID-19

Introduction

At the end of December 2019, a new virus outbreak dubbed by scientists as COVID-19 (SARSCOv2) paralyzed global commerce, generating losses in millions in the world. The virus was rapidly disseminated to the Western cities through international travels and tourism, a problem that led governments to cancel international flights while closing the airspace (Sigala, 2020; Rogerson & Baum, 2020). The WHO (World Health Organization) recommended the imposition of strict lockdowns energetically and keeping social distancing as two valid forms of mitigating the pandemic. In a matter of months, the virus wreaked havoc in the world's main economies, grinding the tourism and hospitality industries to an unparalleled halt. At the same time, each nation promoted and adopted different measures and steps according to their resources and their health systems' operational conditions. Apart from this controversy, more importantly, a whole portion of economies on the planet was slumped down, ushering the capitalist system to one of its worst crises in history. The socio-economic effects of COVID-19 in the tourism industry are manifold and complex and denote a multifaceted dynamic. To some extent, the discipline—which cultivated over years a spirit based on the tourist as the main source of valid information, what A. Franklin dubbed as "tourist-centricity"—is now facing serious problems to study tourism precisely in a

M. E. Korstanje, B. George, *Mobility and Globalization in the Aftermath of COVID-19*,
https://doi.org/10.1007/978-3-030-78845-2_6

(new) world without tourists. While John Urry spoke of the *tourist gaze* as the motivational dynamic of a global (if not abstract) capitalism, the pandemic started the pace for a new feudalized globe characterized by the *wicked gaze*. In the pre-pandemic days, tourists were valorized as ambassadors of "civilization", real global knights widely admired and objects of curiosity, while nowadays tourists are negatively seen as potential enemies or carriers of a mortal disease, widely feared and demonized as never before. In this context, this chapter discusses conceptually the effects and changes of the tourism industry in Argentina in particular but with some broad strokes that illustrate the ways the industry is changing in the world.

As of December 15, 2020, COVID-19 had infected 73,329,034 persons and resulted in 1,630,867 deaths. At first glance, the most affected countries were the US (308,091 deaths), Brazil (181,945), India (143,746), Mexico (114,298), the UK (64,402), France (58,282), Italy (65,011), Spain (48,013), and Argentina (41,041), to name a few. Of course, these numbers keep changing on a daily basis. Argentina, a Southern Latin American country, which adopted a stricter and longer lockdown in the beginning, gradually escalated to the first places in the number of infected people (1,524,372). In medical terms, this point suggests that there were no clear protocols or successful models to control the pandemic. At the same time, each nation adopted different measures in consonance with their resources and health system. The socio-economic effects of COVID-19 in the tourism industry are manifold and complex and evince a multifaceted dynamic (Zheng et al., 2020). While some voices have warned about the need to develop more resilient and eco-friendly forms of consumption (Gössling et al., 2020; Prayag, 2020; Prideaux et al., 2020; Sheller, 2020), others have emphasized the need for changing the current research toward multidisciplinary methods (Wen et al., 2020a, 2020b). In this vein, the discipline that over the years cultivated a spirit based on managerial viewpoint as well as a type of tourist centricity, echoing Franklin, which prioritizes what the tourists say as the only source of information, is now in an epistemological crisis. If John Urry spoke of the tourist gaze as a postmodern dynamic culturally enrooted to capitalism, COVID-19 began with a new feudalized territoriality where the tourist gaze sets the pace to a wicked gaze. Here, the term *wicked* associates not only to the evilness but to a process of demonization over tourists as potential carriers of COVID-19.

From time immemorial, humanity began colonizing the unknown space, indexing new economies, cultures, and peoples. Although travels and expeditions played a leading role in Western cultures' configuration without mentioning a broader readership that consumed travel writings as a new literary genre, it is noteworthy that their morphologies change according to time. On the basis of the technological breakthroughs of the epoch, and, of course, according to the means of economic production, societies adventure to travel beyond the limits of imagination (Korstanje, 2020). Historically hotels served as real *secure places*, as cited by John Bowlby (2005), a type of *zone of protection* which separates the *tourist self from the outside*. In the post-pandemic days, hotels are recycled by the authorities as *shelters* to isolate sick people (Korstanje, 2020). In this troubling context, this chapter discusses conceptually the effects and changes of the tourism industry in Argentina in particular but with some broad strokes that illustrate the ways the industry is changing in the world. The current case study is not representative of other universes but surely sheds light on the dichotomies of the virus in a world without a vaccine. Beyond the interests of academicians to mitigate the socio-economic impacts of COVID-19 in the tourism and hospitality industries, this chapter aims to describe what we have dubbed as "the morphology of travels", which means the cosmologies, narratives, and modes that shape and mold travels. COVID-19 is radically shifting our conception of "Otherness" and the ways travels, as rites of passage, are finally performed.

Initial Discussion

The history of the virus and its impact on modern capitalism seems not to be new. As Jude Fernando (2020) notes, viruses like COVID-19 probe human vulnerability, urging societies to fight with an invisible enemy. The outcomes of societies' responses to the pandemic depend on the human agency and the embodiment of the social and metaphysical relations. The *virocene*, in contrast to the *anthropocene*, is a new age of ecological justice where viruses (virogenic activity) mold the new metabolism of capitalism. Human isolation, social distancing, and the closure of borders not only were steps widely adopted by governments but also teetered the tourism industry to the brink of collapse (Gretzel et al., 2020). In this respect, Bob McKercher documented interesting outcomes on the effects of SARS on the Asian markets (McKercher, 2003; McKercher & Chon, 2004), while emerging studies focusing on H1N1 and Ebola have been brought into the foreground (Monterrubio, 2010; Novelli et al., 2018). Centered on futurist speculations, some of them probably evident and real, experts devoted resources and efforts to implement forecasting methods to avoid pandemics in the next years to come (Jamal & Budke, 2020). As a sensitive activity to global risks, tourism plays a double-edged function in the dissemination of pandemics. On the one hand, through personal or face-to-face contact the virus successfully disseminates. The modern means of transport that daily connect densely populated cities in hours are fertile grounds for mass contagion. On the other, the tourism industry facilitates a rapid expansion of the virus easily, but paradoxically, at the same time, it falls a real victim to the pandemic (Nepal, 2020). To stop the pandemic, governments decided to close borders, airspaces, and commercial activities even tourism worldwide in case of health emergencies. Those overcrowded cities which depend upon tourism consumption experienced serious material losses during the imposition of lockdown. This was exactly what faced the industry in the post-COVID-19 days (Mariolis et al., 2020; Polyzos et al., 2020). The effects in the First and Third Worlds were notably different. Most probably, the most illustrative studies focusing on the effects of COVID-19 in the local economies, even the tourism subsector, do not stem from the Global North, which went to write speculative, theoretical, works, but from the Global South, or peripheral economies which were pressed to an alarming position because of their dependency on "global tourism" (Niewiadomski, 2020). The socio-economic crisis generated by COVID-19 varies by country and levels of economic production (Karim et al., 2020; Edwards, 2020). Rogerson and Rogerson (2020) explore the socio-economic impacts of COVID-19 on the African economies. Based on the South Africa case study, they allude to *spaces of vulnerability* that are previously characteristic of the country. The pandemic has made the material asymmetries already present in each economy and society more acute. Those metropolitan zones or areas which exploit tourism as a mechanism of poverty relief are not only very vulnerable but go through the most negative aftermaths of COVID-19. In consonance with this, Gaffney and Eeckels (2020) affirm that those countries depending on tourism economy are more sensitive to the negative effects of a global quarantine or the closure of borders. Tourists are vividly agents of prosperity and economic welfare for some underdeveloped countries but at the same time they are carriers of the virus. This creates a contradictory situation that is very hard to resolve. To expand the understanding of this quandary, Gaffney and Eeckels suggest the enumeration of low- and high-risk countries affected by the pandemic. In consonance with this,

Farzanegan et al. (2020) argue that those nations characterized by high levels of mobilities are prone to experience more deaths and infected cases than low-mobile countries. Those economies dotted with inbound and outbound international travels have serious problems coping with the virus outbreak. The refusal to close airports, borders, and airspace aggravated the situation, accelerating inevitable mass contamination. In this vein, Sah et al. (2020) test these assumptions in Nepal while Korstanje (2020) proffers the case of Argentina where thousands of tourists stranded abroad were expatriated to Buenos Aires. In Nepal, the virus put the health and tourism industry in jeopardy, increasing the health budget to 6% for the next fiscal year (Sah et al., 2020). In Argentina, the tourism industry and the accommodation offer faced. At a closer look, tourists were demonized as potential carriers of the virus whereas chauvinist and racist expressions against neighboring migrants emerged in metropolitan Buenos Aires (Korstanje, 2020). This raises an interesting question, is COVID-19 indicating the end of hospitality as we know it?

The Essence of Hospitality, New Dialogues with Derrida

Over the years, sociologists and anthropologists have been motivated by the study of hospitality (Selwyn, 2000; Lashley, 2008). To wit, social sciences have been moved to analyze hospitality, though from different angles. The notion of hospitality was investigated from a different lens and using different theories but they can be very well grouped into two families. Those works focusing on the anthropological giving while returning (exchange) rite are the basis of cultural reciprocity and those that understand hospitality as an efficient instrument of social control (Lynch et al., 2011). At a closer look, hospitality corresponds with an inter-tribal institution of protection and exchange in peacetimes or warfare. Because of this, the foreigner should be protected and respected in the same ways God protects human souls in the afterlife. In fact, hospitality occupies a central position in the configuration of Western rationality but also in the layout of the modern nation-state (Korstanje, 2020). Having said this, hospitality can be offered or negated to the stranger. Today, millions of migrants suffer the strict controls and surveillance technology imposed to check them at borders, citing Bauman, as real vagabonds of a global world. Late capitalism rests on a great contradiction. While tourists are considered heroes who evince Western capitalism's triumph, migrants are the victims, probably immobilized to remain in the periphery (Lyon & Bauman, 2013; Bauman, 2017). This aspect aligns with a great controversy revolving around the mobilities problem and the structural inequalities accelerated by tourism consumption (Tzanelli & Korstanje, 2020; Jayathilaka, 2020). As Daniel Innerarity (2017) explains, outsiders, who do not look like us, interrogate further the local customs and institutions invoking the sacred laws of hospitality. Those societies which neglect hospitality are culturally doomed to disappear. This happens because hospitality is the touchstone of human reciprocity, as Innerarity contends. In his book, *Terrorism Tourism and the End of Hospitality in the West*, Korstanje (2017) holds the opinion that 9/11 ignited a process of social decomposition, probably long dormant in the Western social imaginary. Terrorism ignited a new epoch widely inscribed in the fear of the stranger, the "non-Western Other", not only affecting the position of the West with respect to the rest of the world but marking the end of hospitality. Terrorism closed and mined West from inside, destroying its most important hallmark, *the sacred law of hospitality*. Similar concerns about the end of hospitality can be found in Ritzer (2019) and Selwyn (2019). Both authors refer to the rise of anti-foreign or anti-migration policies without mentioning the appearance of chauvinist and separatist discourses as a clear sign showing hospitality is in crisis. Starting from the premise that the globalizing process has blurred the borders stimulating travels and migratory fluxes across the globe, this does not authorize to say that the world is global. Countless anti-globalization movements have emerged in recent years. All these romantic movements echo the same sentiment, centered on an unquestionable hostility against the stranger. Ranging from White supremacist discourse embodied after Donald Trump's triumph in the US to the Brexit or the Windrush scandal in the UK, the established Western democracies debate between liberty and the fear of the stranger. Ritzer (2019) and Selwyn (2019) acknowledge that the alterity is seen as a potential enemy in this new era, which disrupts the social order. In this book, we have delved into Derrida on several occasions. From different angles, we wanted to present a coherent viewpoint of Derrida, who has been historically misunderstood, in particular his notion of hospitality. Next, we give a fresh snapshot of the intersection between *the guest*, who is always desired, and *the parasite* persecuted and neglected.

Of Hospitality

Without any doubt, hospitality has been academically discussed within the constellations of social science, but in its nature, the term remains misunderstood. Jacques Derrida is one of the pioneering voices who have focused their attention on hospitality. However, his legacy—for reasons which are hard to explain here—has been correctly caught by the academic circles. To cover the caveat, we will debate the legacy left by Derrida and his contributions and limitations on hospitality. Although Derrida and Dufourmantelle (2000) has published countless texts about hospitality and migration, the point is brilliantly captured in his book entitled *Hospitality* in which he bolsters a fluid dialogue with Anne Dufourmantelle debating the problem of the "Otherness" in the Western lifestyle. The selfhood is formed in opposition to the presence of alterity, which acts as a mirror reflecting what the selfhood represses. As a part of a dark (repressed) side, the "Other" bursts into the room without knocking. Organized in two lectures, the book dissects how the alterity invariably dissolves in the hegemony of language. To wit, the alien defies not only the "reign of dogmatism", as Derrida alerts, but also the essence of the law. In his classic text, *Sophist*, Plato calls attention to the role played by Xenos in philosophy. The stranger (Xenos) should be conceptually distinguished from the foreigner. While the former refers to those who are accompanied by a different language than we normally speak, the latter interrogates further the hosting community re-elaborating the societal institutions. Starting from the premise that these outsiders threaten the paternal hegemony, it is tempting to say that strangers are placed on the lens of strict scrutiny. The apparatuses of states originally designed to ensure the social order often exert violence on the guest's body when it becomes a disruptive agent. Not surprisingly, hostility and hospitality share the same etymological origin. Hostility, against aliens, activates the restitution of lost borders, reviving a sentiment of radical separation with the "Other".

Hospitality, like language, is hosted by the question. Nobody offers hospitality to a stranger or a person without an identity, just as nobody would seek hospitality from unidentifiable businesses. When the guest asks for shelter, the hosting community asks: who are you and what do you want? (Almeyda-Ibáñez & George, 2017). Derrida adds that heritage and economic patrimony are two essential factors of hospitality. Philosophically speaking, the stranger splits our phenomenological (sensible) world into two parts, us-living-here and them-living-there. It is common to think that the identity derives from our family, kinship, city, or nation; nonetheless, the notion of identity constitutes the "Otherness" position. The "Other" who does not speak my language, as an outsider, not only knows me but sees aspects of my life that remain inexpugnable to me. Any question, like the guest, can be formulated (hosted) or rejected.

If the question of language situated a central position in Derrida, "the foreigner's rights" follows a similar dynamic. The "Other" is often admired as an outsider (a tourist) who can pay for hospitality whereas the foreigner is labeled as an external enemy (migrant). Nobody offers hospitality to people who lack a name and probably this is one of the most notable paradoxes of hospitality. The question of hospitality associates with heritage (and history). An exiled citizen has no past, no name, nor heritage. There are a lot of stories and myths in ancient cultures where the figure of exile equates to a symbolic, if not real, death. This happens simply because anonymity lies excluded from hospitality. After all, nobody offers lodge to a person who is not recognized by the language. The divorce of the "Other" and the "Stranger" as different categories is vital to understand the difference between absolute and conditioned hospitality. Absolute hospitality is based on a utopia where hosts ask nothing in return from the guest, whereas conditioned hospitality is offered for those who can pay for it. Here two assumptions are to be made. Tourists pay for the service they receive and, of course, they can be considered as global "Others" while the working migrant equates to the figure of the "stranger". If tourists need medical assistance, their insurance covers the cost, but the migrants have no such luck. Hospitals certainly are examples of the spirit of absolute hospitality, while conditioned hospitality is the quintessential feature of hotels. Let's clarify to readers this argumentation is important to understand the remaining sections. But coming back to Derrida as the main line of inquiry, it is argued that foreigners are systematically subject to the local law just upon arrival. The right which the family, society, and state frame has temporarily bestowed the alien has to be protected, but once the law is broken, hostility arises. At a hotel, which is part of the private sphere, the host and the guest interact freely, but if a crime is committed, the police intervene in the scene interrogating the stranger. At this moment, hospitality temporarily disappears. What is equally important, hospitality invites us to break the law by establishing authorities and moral limits to our actions. Last but not least, for regulating the host-guest interaction, hostility marks the pace for distinguishing (identify) the *desired* from the *undesired guest* (a parasite). Here, we have tried to offer a clear and short viewpoint of Derrida and his argument revolving around hospitality. Is COVID-19 an undesired guest?

At first glance, biological sciences have amply evinced that the function of viruses links to strengthen populations. The virus infects an organism, replicating itself during a period, leaving the organism to be hosted by "others". This silent enemy remains inexpugnable for the human eyes, sometimes undermining human omnipotence and "the ocularcentrism". Dixon and Jones analyze the dichotomies of "tactile topologies" to connect with mobile bodies. Centered on the film *Contagion* (2011), they call attention to the rise of new narratives oriented toward demonizing "the face to face contact with the Other". In Soderbergh's *Contagion*, so to speak, they infer that human geographies are orchestrated to draw an intimate space where social contact makes people more vulnerable. They go on to say:

> In dwelling upon such objects, Contagion draws us into the intimate, affective space of the micro, teeming with life and demonic energy, of which we partake through touch. To touch in this environment is to position the skin not as a superficial, or exterior, container to the flesh, but as a porous membrane, producing its own complex geographies fuelled by a world of distributed objects and the two to three thousand hand-to-face touches we make each day. Through touch, the body becomes vulnerable to the potentially pathological and not-so-distant other—putting us at constant risk of being penetrated, invaded and over-run by the micro's ever-proliferating inhabitants. (Dixon & Jones, 2015: 230)

In consonance with Dixon and Jones, we strongly emphasize the metaphors of hospitality, as least as Derrida imagined it, to help understand how travels and human geographies are being abruptly changed.

The Virus as an Undesired "Other"

In earlier texts, we have commented that hospitality as the sacred law that erected Western civilization's legitimacy and expansion is dying after 9/11. Terrorism opened the doors to a new epoch where the "Other" has become an "undesired guest". The colonial period, endorsed by the imperial expansion of European nations worldwide, cultivated a paternalist taste for the "stranger", who as a "non-Western Other" would be re-educated according to the European cultural values. The aborigines not only occupied a central position in travel writing but were the object of curiosity and fear for Europeans. Travel and expeditions to overseas territories were oriented toward strengthening the center-periphery dependency while proffering cultural entertainment for European audiences. The need for discovering new lands, new customs, or simply "being there" was prone to colonial expansion. Europe encouraged locals to import hospitality as a paternalist form of relationship with the "Other". This does not mean that this "Other" was accepted without any qualms. The alterity was drawn as an inferior being respecting the European superman. Aborigines had to be re-socialized according to the European mainstream cultural values for them not to perish. With 9/11, this romantic view has substantially changed. The "Other" has become a potential enemy who emulates living in the society he hates. Based on the doctrine of living with the enemy, this emerging discourse paved the way for a more radicalized discourse where the "non-Western Other"—even in the case of those who look like us—is a silent killer looking for the opportunity to perpetuate the next attack (Korstanje, 2017, 2018). Terrorism finally closed the doors of heaven. The recent COVID-19 outbreak, far from being a foundational event as some scholars admit, aligns with the end of hospitality, which is a process started in the post-9/11 context. Today the war on terror sets the pace for the war against a virus.

In the following section, we will give some explanations on the feudalizing process Argentina is suffering because of COVID-19. Next, we debate the effects of COVID-19 accelerating the crisis of the tourism industry and the decline of hospitality given three different lenses: the "Other" as an undesired guest, the inevitable feudalization process which separates states, provinces, and counties, and finally the precautionary logic inscribed in the pay today for consuming tomorrow's logic. As stated in the introductory section, all these dynamics are gradually shifting the morphology of travels, as least as we know it.

The Undesired Guest

On March 19, 2020, moved by the pandemic's proliferation, President Alberto Fernandez imposed a strict mandatory lockdown that included the closure of borders and airspace, the cancelation of flights, accompanied by the closure of schools, universities, and public edifices. The mandatory lockdown generated serious economic impacts on the commerce and tourism industry. To placate the quarantine's negative effects, the government began to bail out companies in bankruptcy. À la Derrida, hotels, which were seen as emblems of conditioned hospitality, were refurbished into hospitals to assist infected people with mild symptoms. In the days after the total closure, thousands of Argentinean citizens (tourists) were stranded abroad. Meanwhile, countless foreigner tourists gallivanting around Buenos Aires streets were in search of solutions for their repatriation. The world, at least as we know it, was put upside down. Foreigner tourists and expatriates were marked as potential carriers of the virus while the police's efforts were certainly put in prohibiting the essence of the nation-state, so to speak, the free circulation. The first point of entry in this discussion is the figure of the tourist as an ***undesired guest***. Police and security forces, often escorted by migration officers, were dispatched to hotels to find, imprison, and exile those foreign tourists who have violated the quarantine or abandoned their hotels. These security protocols were followed by different legal sanctions and penalties that applied to those domestic tourists who drove toward second homes in the hinterland. Doubtless, citizens were strictly scrutinized or subjected to an immobile climate of control, as never before. In particular, their right to travel was temporarily cancelled by the state. The inter-state borders were closed, isolating states from free trade. The question of whether John Urry coined the term ***tourist gaze*** to denote the rise of a new form of consumption based on ocularcentrism sets the pace for a new conception, where gazing was simply banned. This was called a "wicked gaze" (Korstanje, 2020). The "Other" is no more an object of curiosity in these new times; rather, it is a source that inspires hostility and fear. The process explains the feudalization process or radical transformation of national border capitalism in post-COVID-19 days. The discourse of living with the enemy inside, which was inaugurated after 9/11, remains alive.

Re-feudalization Process

A re-feudalization process is defined as a political condition of desegregation where units act with autonomy respecting the former central administration. The word stems from the feudalization process initiated in Europe just after the collapse of the Roman Empire where European feuds interacted with their neighbors with autonomy from Rome. The modern nation-state's appearance has historically lumped together many feuds and ethnic groups into the idea of a common heritage dubbed as "nationhood". Nation-states reserved and centralized the use of force and police through the orchestration of legal mechanisms of control such as customs, constitution, and national borders. This political unit, which is known as nation-state, administered the relation of different sub-units (provinces or states) (Guidotti-Hernández, 2011). One of the effects of this pandemic interlinks to the fragmentation of some provinces in regard to the decision made in the central administration, without mentioning the emerging disputes between the federal government and governors or majors. These types of crises seem to be probable when the solidarity, if not the legal liaison, among states or provinces is undermined. If the global era was characterized by the erosion of some national borders, COVID-19 started a new re-feudalization process marked by separatism, racism, and the re-affirmation of internal borders. This goes from barricades to impede the entrance of outsiders to the unilateral imposition of migration-related requirements (documents) which sound at least unconstitutional.

Lastly, Argentina in post-Covid-19 period witnessed a chaotic situation where its 24 provinces did not unilaterally follow health protocols. Buenos Aires and the metropolitan area, known as AMBA, were in strict lockdown, closing restaurants, malls, hotels, and commercial activity, while other cities such as Jujuy, Rosario, Mendoza, and Cordoba entered a "new normality". In the following months, the virus dissemination compelled authorities to reinforce restrictive measures, passing from the new normality to a strict quarantine stage. The second element of this process was the *feudalization of mobilities.* The country was broken into pieces and the central administration failed to impose a planned program to regulate frontiers. Each province (state) not only reserved the rights of police but, surely contradicting the constitution, each one regulated the borders at their discretion. States such as San Luis, San Juan, and Mendoza banned the entrance of other Argentinean tourists and travelers. Only food transport, which was tightly scrutinized, was permitted to enter and leave the state during the day. The Province of Formosa remains impermeable to receiving travelers, no matter their profession, even to date. Some lawsuits were presented before the Supreme Court of Justice because of the unconstitutionality of travel ban local dispositions. Hundreds of Argentinean tourists, drivers, and travelers were stranded in the provinces for months without assistance, shelter, or food. In other cases, drivers were subject to strict protocols of control and surveillance. In a third stage, some states opened their frontiers allowing other inter-state visitors but they were obliged to show different documents such as certificates of COVID-19 tests, adjoined to health coverage or the purchasing of insurance protection while traveling, without mentioning the submission of a sworn administrative form. This feudalization logic was strictly based on ever-enlarging autonomies of states respecting the central administration. Basically, in the pre-pandemic contexts, tourists—no matter what their nationalities—were welcomed as "friendly visitors" who contributed to boosting local economies. In the post-pandemic days, these guests have become *undesired guests* who are potential carriers of a mortal disease.

Pay Today While Traveling Tomorrow

The third element to discuss is the "pre-travel campaign" encouraged by the tourism ministry to promote tourism consumption at the end of lockdown. At the end of October 2020, the tourism ministry launched a national campaign to stimulate inter-state travel. The economic program included a set of economic incentives, awards, and financial assistance to tourists to buy now—in 2020—to receive discounts and bonuses in 2021. The *pre-travel* consisted of a program of tourism presales oriented toward crediting economic benefits in 2021. Registering on a specific website while loading the purchasing bills, tourists received 50% plus in credits to be redeemed at the destinations once they arrived. Visitors were encouraged to register and provide their personal information until October 30 whereas a credit card issued by the central bank allowed tourists to buy for 50% of the incurred expenditure. To provide an example, a family group buys through a travel agency a tour package which includes air tickets for four persons, hotel, and excursion totaling ARS 150,000. This prepaid tour is for Calafate to be used in February 2021. Meanwhile, two air tickets to Cordoba are added to the package for ARS 10,000 to be redeemed in August 2021. The invoices are loaded to the webpage of the tourism ministry for a total of ARS 160,000, in which case, the family receives a credit of ARS 80,000. Soliciting visitors receive a (virtual) electronic wallet for ARS 75,000 to be used in Calafate and ARS 5,000 in Cordoba since August. The economic program fostered by the tourism ministry seems not to be new, since it was applied worldwide, and centered on the logic of paying now and consuming later. What is equally important, the campaign to travel in the coming year inscribed a futurist scenario where the precautionary doctrine prevails. With the benefit of hindsight, the precautionary doctrine operates from a hypothetical future marking the pace in the present. Of course, all-inclusive tours follow the same dynamic, purchasing now what should be consumed later. The difference in the post-COVID-19 context associates to discounts, promotions, and benefits invested by the state to encourage tourism consumption. In some cases, authorities who are representative of local tourism ministries evaluated the option to apply taxes to outsiders to keep the health system working.

As stated, tourism plays an ambiguous role as a growing industry, which is a fertile ground toward dissemination, but at the same time, it helps in the economic recovery of society (George & Ramirez, 2019). This recovery facet is accelerated in those economies where tourism is re-appropriated as an instrument toward an economic revitalization. Tourism seems to be part of the problem of mass consumption and higher mobilities, accused by experts as the reasons for the virus contagion, but at the same time, the resiliency of the industry proves to be a hope for a rapid recovery. In the same way, at a preliminary stage, governments' efforts were needed to control the tourist fluxes, but now the industry has opportunities for rebirth from the dawn of disaster. Those readers interested in tourism history realize that travels in the earlier centuries were carried out by rich people who took advantage of the summer season to move to beaches and holiday centers. A large proportion of the workers not only had no paid holidays but lacked the opportunities to travel. The working benefits fostered by worker unions along with the technological advances paved the way for the rise of what historians know as "mass tourism".

In the post-COVID-19 context, mass tourism goes through a new facet of atomization where only a limited number of persons travel. The ruling elite, citing Bauman, is legally and economically able to travel and enjoy the benefits of tourism while a large percentage of the workforce is forced to stay at home. Stay at home is a mandate dictated by the state to mark those who are so special as they want.

Conclusion

Although the history of the virus and its impact on humanity are not new, nobody anticipated the effects of COVID-19 on the global trade and sub-service sectors. The tourism and hospitality industries are not an exception. The pandemic paralyzed these growing activities as well as the global trade. Theorists debated the importance of implementing a specific agenda to mitigate the negative socio-economic effects of COVID-19 in the industry. While some experts called for a new opportunity to redeem a more ecological form of consumption, others alluded to elaborating emergent methodologies that help reconsider tourism's nature.

With the focus on Argentina as a case study, this chapter reviewed the ebbs and flows of the government in implementing restrictive measures organized to stop human mobilities and travels. We found three relevant aspects of the pandemic. The undesired guest marks the pace for a new understanding of foreigner tourists that equate to the expatriates who are subjects of mistrust and fear. A feudalization process marks a clear geographical separation among provinces regarding the central administration. The classic cosmologies of travels—based on belonging to a shared territory and heritage—has mutated to a decentralized unity where each province reserves the right to reject or accept other Argentine travelers. The term *visa* is derived from the Latin word *visum*, the verbal past tense of *videre*, which means "to see". Citing Derrida, this suggests that the nation endorses the right of hospitality only if the identity of the visitor can be tested. The fact is that nation-states reserved the right to interrogate further on the stranger's identity. Even if strictly prohibited by the constitution, now this right has been ceded to the provinces. The third element associates with the program pay now while consuming later, a program planned by the tourism ministry to stimulate the industry. As stated, the present case study, far from being applied to other universes, can be understood in the context of Argentina. This chapter aims at describing the new cosmologies of travels and tourism which passed from an unrestricted liberality to a technocratic control. We have coined here the term *technocratic travels* to denote the use of digital technologies to monitor and scrutinize while putting serious barriers to human mobilities. In a technocracy, the system of governance is strictly based on the use of technology. Technical knowledge regulates not only human interaction but also the decision-making process in governments. The technocracy contrasts sharply with the deliberative democracy which places the human mind in the center of the scene. In a technocratic society, knowledge production is drawn by bureaucracy and depersonalization. COVID-19 paves the way for the rise of a technocratic society where standardized and dehumanized protocols regulate travels; in other words, we pass from leisure to technocratic travel. For moving through the national territory, Argentineans need to present countless documents and technocratic qualifications as never before.

References

Almeyda-Ibáñez, M., & George, B. P. (2017). The evolution of destination branding: A review of branding literature in tourism. *Journal of Tourism, Heritage & Services Marketing, 3*(1), 9–17.

Bauman, Z. (2017). Tourists and vagabonds: Or, living in postmodern times. In *Identity and social change* (pp. 19–32). Routledge.

Bowlby, J. (2005). *A secure base: Clinical applications of attachment theory* (Vol. 393). Taylor & Francis.

Derrida, J., & Dufourmantelle, A. (2000). *Of hospitality*. Stanford University Press.

Dixon, D. P., & Jones, J. P., III. (2015). The tactile topologies of Contagion. *Transactions of the Institute of British Geographers, 40*(2), 223–234.

Edwards, R. B. (2020). Bubble in, bubble out: Lessons for the COVID-19 recovery and future crises from the Pacific. *World Development, 135*, 105072.

Farzanegan, M. R., Gholipour, H. F., Feizi, M., Nunkoo, R., & Andargoli, A. E. (2020). International tourism and outbreak of coronavirus (COVID-19): A cross-country analysis. *Journal of Travel Research*. 0047287520931593.

Fernando, J. L. (2020). The Virocene Epoch: The vulnerability nexus of viruses, capitalism and racism. *Journal of Political Ecology, 27*(1), 635–684.

Gaffney, C., & Eeckels, B. (2020). COVID-19 and tourism risk in the Americas. *Journal of Latin American Geography, 19*(3), 308–313.

George, B., & Ramirez, A. (2019). Plastic recycling and waste reduction in the hospitality industry. *Economics, Management and Sustainability, 1*, 6–20.

Gössling, S., Scott, D., & Hall, C. M. (2020). Pandemics, tourism and global change: A rapid assessment of COVID-19. *Journal of Sustainable Tourism, 29*(1), 1–20.

Gretzel, U., Fuchs, M., Baggio, R., Hoepken, W., Law, R., Neidhardt, J., ... & Xiang, Z. (2020). e-Tourism beyond COVID-19: A call for transformative research. *Information Technology & Tourism, 14*(1), 3–19.

Guidotti-Hernández, N. M. (2011). Unspeakable violence. In *Remapping US and Mexican national imaginaries*. UNCS Press.

Innerarity, D. (2017). *Ethics of hospitality*. Taylor & Francis.

Jamal, T., & Budke, C. (2020). Tourism in a world with pandemics: Local-global responsibility and action. *Journal of Tourism Futures, 6*(2), 181–188.

Jayathilaka, G. K. (2020). The worldmaking agency of the Sri Lankan travel blogger. *Tourism Culture & Communication, 20*(2–3), 117–127.

Karim, W., Haque, A., Anis, Z., & Ulfy, M. A. (2020). The movement control order (mco) for covid-19 crisis and its impact on tourism and hospitality sector in Malaysia. *International Tourism and Hospitality Journal, 3*(2), 1–7.

Korstanje, M. E. (2017). *Tourism, terrorism and the end of hospitality in the West*. Palgrave Macmillan.

Korstanje, M. E. (2018). *The challenges of democracy in the War on terror: The liberal state before the advance of terrorism*. Routledge.

Korstanje, M. E. (2020). Passage from the tourist-gaze to the wicked-gaze: A case study on COVID19 with special reference to Argentina. In B. George & Qamar-ud-Din Mahar (Eds.). *International case studies in the management of disasters: Natural–manmade calamities and pandemics*. Emerald Group Publishing.

Lashley, C. (2008). Studying hospitality: Insights from social sciences. *Scandinavian Journal of Hospitality and Tourism, 8*(1), 69–84.

Lynch, P., Germann-Molz, J., McInthosh, A., Lugosi, P., & Lashley, C. (2011). Theorizing hospitality. *Hospitality & Society, 1*(1), 3–24.

Lyon, D., & Bauman, Z. (2013). *Liquid surveillance: A conversation*. Polity.

Mariolis, T., Rodousakis, N., & Soklis, G. (2020). The COVID-19 multiplier effects of tourism on the Greek economy. *Tourism Economics*, 1354816620946547.

McKercher, B. (2003). SIP (Sars induced Panic) a greater threat to Tourism than SARS (Severe acute respiratory syndrome). *E-revise of Tourism Research (ERTR), 1*(1), 17–18.

McKercher, B., & Chon, K. (2004). The over-reaction to SARS and the collapse of Asian tourism. *Annals of Tourism Research, 31*(3), 716–730.

Monterrubio, J. C. (2010). Short-term economic impacts of influenza A (H1N1) and government reaction on the Mexican tourism industry: An analysis of the media. *International Journal of Tourism Policy, 3*(1), 1–15.

Nepal, S. K. (2020). Travel and tourism after COVID-19–business as usual or opportunity to reset? *Tourism Geographies, 22*(3), 1–5.

Niewiadomski, P. (2020). COVID-19: From temporary de-globalisation to a rediscovery of tourism? *Tourism Geographies, 22*(3), 1–6.

Novelli, M., Burgess, L. G., Jones, A., & Ritchie, B. W. (2018). No Ebola… still doomed'—The Ebola-induced tourism crisis. *Annals of Tourism Research, 70*, 76–87.

Polyzos, S., Samitas, A., & Spyridou, A. E. (2020). Tourism demand and the COVID-19 pandemic: An LSTM approach. *Tourism Recreation Research, 46*(2), 1–13.

Prayag, G. (2020). Time for reset? COVID-19 and tourism resilience. *Tourism Review International, 24*(2–3), 179–184.

Prideaux, B., Thompson, M., & Pabel, A. (2020). Lessons from COVID-19 can prepare global tourism for the economic transformation needed to combat climate change. *Tourism Geographies, 22*(1), 1–12.

Ritzer, G. (2019) Inhospitable hospitality. In B. Rowson & C. Lashley (Eds.), *Experiencing hospitality* (pp. 73–91). Nova Science.

Rogerson, C. M., & Baum, T. (2020). COVID-19 and African tourism research agendas. *Development Southern Africa, 13*, 1–15.

Rogerson, C. M., & Rogerson, J. M. (2020). COVID-19 and tourism spaces of vulnerability in South Africa. *African Journal of Hospitality, Tourism and Leisure, 9*(4), 382–401.

Sah, R., Sigdel, S., Ozaki, A., Kotera, Y., Bhandari, D., Regmi, P., ... Dhama, K. (2020). Impact of COVID-19 on tourism in Nepal. *Journal of Travel Medicine, 27*(6), taaa105.

Selwyn, T. (2000). An anthropology of hospitality. In C. Lashley & A. Morrison (Eds.), *In search of hospitality: Theoretical perspectives and debates* (pp. 18–37). Butterworth-Heinemann.

Selwyn, T. (2019). Hostility and hospitality: Connecting Brexit, Grenfell and Windrush. In B. Rowson & C. Lashley (Eds.), *Experiencing hospitality* (pp. 51–72). Nova Science.

Sheller, M. (2020). Reconstructing tourism in the Caribbean: Connecting pandemic recovery, climate resilience and sustainable tourism through mobility justice. *Journal of Sustainable Tourism, 29*(2), 1–14.

Sigala, M. (2020). Tourism and COVID-19: Impacts and implications for advancing and resetting industry and research. *Journal of Business Research, 117*, 312–321.

Tzanelli, R., & Korstanje, M. (2020). Introduction: Critical thinking in tourism studies. *Tourism Culture & Communication, 20*(2–3), 59–69.

Wen, J., Wang, C. C., & Kozak, M. (2020a). Post-COVID-19 Chinese domestic tourism market recovery: Potential influence of traditional Chinese medicine on tourist behaviour. *Anatolia, 32*(1), 1–5.

Wen, J., Wang, W., Kozak, M., Liu, X., & Hou, H. (2020b). Many brains are better than one: The importance of interdisciplinary studies on COVID-19 in and beyond tourism. *Tourism Recreation Research, 97*, 1–4.

Zheng, Y., Goh, E., & Wen, J. (2020). The effects of misleading media reports about COVID-19 on Chinese tourists' mental health: A perspective article. *Anatolia, 31*(2), 337–340.

CHAPTER 7

COVID-19, Terrorism, and Conspiracy Plot: Everyone Is Guilty Until Proven Innocent

Introduction

The recent virus outbreak at the end of 2019 paralyzed China and the world in a few months. SARSCOv2 (known as COVID-19) appeared first in Wuhan city, China, and rapidly disseminated to Europe, the US, and Latin America. As of December 15, 2020, COVID-19 had infected 73,329,034 persons and resulted in 1,630,867 deaths. At first glance, the most affected countries were the US (308,091 deaths), Brazil (181,945), India (143,746), Mexico (114,298), the UK (64,402), France (58,282), Italy (65,011), Spain (48,013), and Argentina (41,041), to name a few. These numbers are constantly changing. Given this mayhem, Western governments adopted restrictive measures to limit public circulation and mobilities. The measures consisted of the closure of the airspace, borders, and international tourism, accompanied by other actions that paralyzed global commerce. Some countries sanctioned a curfew to avoid public circulation. Doubtless, COVID-19, at least for the media and social imaginary, situated as a turning point, or re-foundational event, that awoke the capitalist world from the slumber it was in (Zizek, 2020). Terms such as social distancing, new normality, and lockdown resonated in public opinion, buttressing long-dormant and critical discourses against capitalism which vaticinated the beginning of a new, more sustainable capitalism (Laster Pirtle, 2020; Springer, 2020). However, beyond this enthusiasm,

M. E. Korstanje, B. George, *Mobility and Globalization in the Aftermath of COVID-19*,
https://doi.org/10.1007/978-3-030-78845-2_7

this chapter brings a deep reflection on the fact that, far from being a foundational event, COVID-19 is a previously established process initiated after 9/11 which we dubbed as "the end of hospitality" (Korstanje, 2017, 2019). The attacks on New York City in 2001 surely accelerated the emergence of a new epoch, where the classic image of the "non-Western Other" as irrational, primitive, uneducated, and so to speak a child sets the pace for a more radicalized view embedded in a much deeper sense of securitization. This tragedy was the first time four airplanes were weaponized against the most important civilian targets of the US. The so-called war on terror described an unprecedented situation where the "enemy" did not come from the outside; now it resides with us, it looks like us, and it may even be a neighbor or a friend. The conspiracy narratives occupied a central position in the configuration of the post-9/11 context, aligning with the idea of fighting against an enemy very hard to locate and eradicate. A long-dormant culture of fear allowed the rise of radicalized discourses, and even to Donald Trump's access to the presidency of the US, but what is more important is that it accelerated a moral crisis that closed Western civilization to the alterity (Korstanje, 2017).

Of Hospitality?

In recent years, the problem of terrorism, as well as the fear of mass migration, has occupied a central position in the public debate, not only setting the agenda in the media but also igniting some radicalized discourses in the consolidated Western democracies (Bauman & Lyon, 2013; Howie, 2012; Skoll, 2016; Altheide, 2017; Korstanje, 2017). The Brexit, Windrush scandal, the Wall in Trump's speech, tourist-phobia, and Islamophobia without mentioning the conflict in Gaza narrate a daily situation where the "Other", at least the "non-Western Other", is not the desired guest anymore (Allen, 2010; Milano et al., 2019; Ritzer, 2019; Selwyn, 2019). As David Altheide puts it, the culture of fear, associated with the problems in understanding alterity, paved the way for a new regime where the democratic institutions set the pace for more radicalized discourses and sentiments (Altheide, 2017). This sentiment invariably invites us to rethink the conceptual pillars of hospitality as the cultural core of Western civilization and, of course, the efforts of terrorism to dismantle hospitality from within (Korstanje, 2017). In so doing, broadly speaking, we need to trace back the roots of hospitality to Derrida's texts, in particular, his main argumentation exhibited in his book *Of Hospitality*, an interesting dialogue with Anne Dufourmantelle, which was originally published in English by Stanford University Press in 2000.

As a backdrop, Derrida starts from the premise that the problem of hospitality cannot be grasped without touching the notion of Xenos (the foreigner) whose presence interrogates further the nativity. As he eloquently infers, Plato described the alien (Xenos) as somebody who asks about us shaking the kingdom of "dogmatism". However, unlike the king, the foreigner does not belong to the hosting culture he finally interrogates. The guest seems to be always accompanied by a foreign language and culture, threatening the paternal hegemony. Consequently, foreigners are simply strangers who do not speak our language and ultimately do not share "our common understanding". Foreigners are always suspected to be potential enemies unless they prove to be non-guilty. The question of whether the stranger divides our world into two parts speaks to us of the fact that hospitality can be given or rejected. To put this simply, strangers often know more about our customs, habits, and laws than we think. Everyone understands the world through the lens of "Others", and of course, Western civilization is not an exemption. Here, Derrida revolves around the limitations of identity as a phenomenological experience situated in the "Other". Lay-people certainly acknowledge that identity comes from our parents, laws, and traditions, but indeed it derives from the presence of the alterity and can only be recovered through the Other's lens. Ethnocentrism packages and covers a set of repressed feelings the aliens can unpack. This means that outsiders often see things we repress or simply the truth we do not want to accept, at least overtly. In Derrida's terms, the guest-host encounter is essentially troublesome, like a philosophical question. He argues convincingly that the question, like the foreigner, holds in the language and very well can be formulated or negated. The welcoming process accepts the notion of risk because the "Other" remains unknown in its intentions. This moot point begs a more interesting question: to what extent is it possible to interrogate when hosts are not accepted in the first place?

Without any doubt, the above-formulated question embroils a sophisticated metaphor. A clear philosophical dilemma is expressed when a person is out of the home. Citing Socrates' mediations, Derrida explains that his main error was not to be familiar with judges' language. Socrates not only rejected but misunderstood the "language of the trial", inferring that he was ultimately declared guilty without a just defense. Socrates never shared the same linguistic code as his judges and faced the fury of the

Court "as a foreigner". Frequently, the host's language interpellates violently to aliens while forcing them to adopt a singled out interpretation. The possibilities of hospitality take place at the moment the unknown words are incorrectly translated. The translation appears to be a passage, a type of rite of domestication where the "unknown and there" is internalized as the known and here. Nobody offers hospitality to travelers who cannot attest their identities. As Derrida observes, the notion of hospitality associates strictly with the power of the name, which denotes a patrimony, a city of origin, or a family. Those persons who lack a name are not only exiled but also demonized. The figure of identity plays a crucial role in the distinction between absolute and restrictive hospitalities. While the former signals a utopian situation where the "Other" is hosted without asking anything in return, the latter connotes the plausibility to be offered only when the guest can pay for it. The restrictive hospitality asks directly, who are you and what do you want?

Centered on Hegel's thesis, Derrida understands that hospitality confers rights to strangers simply because once they arrive in a country they are immediately subject to local laws even if they do not know them. Constructed through a communal ethos, foreigners receive certain protection only under the condition that the established status quo is not defied. There is a liaison between hospitality and hostility, which share the same origin. What is equally important, at a hotel or in a shopping mall the guest can interact with the host privately, without any state intervention. But things are radically changed when a person is harmed or killed. The crime opens the doors to the suspension of intimacy, as well as the autonomy of guests. Once the police take over the scene, the actors are interrogated or their phone lines tapped, suspending temporary hospitality. The structures of the state, law, and justice discipline the alien when crime emerges. In such a moment hospitality sets the pace for hostility. Not surprisingly, thousands of migrants are pursued, jailed, and exiled when they flout the rules. This happens because the relationship with a foreigner is ruled by the right to be part of the correct side of justice. As Derrida discusses, hospitality calls for a paradoxical situation. On the one hand, it invites to break the rules affirming the authority of the state to correct things, as the safeguard of social order. On the other, it transgresses the conception of universal citizenship, à la Kant. In this vein, Derrida echoes the Oedipus legend to bring his point to the foreground. Oedipus is self-exiled by his crime, disposed of any right or any law. Antigone, his daughter, wants to look for her father's corpse, but she is not aware where he passed away. It seems as though he liked to depart, to be exiled, and he left some traces and hints to guide his daughter to find the corpse. He, after all, does not leave this earth without a plan so that he can be remembered eternally, but what is more significant is that Antigone commits suicide because she is unable to locate and repatriate Oedipus's corpse. It is important not to lose sight that Oedipus's myth clearly explains the philosophical quandary of foreigners. The host allows the foreigner to be protected and lodged at his home, of course after receiving a formal invitation, but once done, his guest controls him by manipulating secrecy. All appears to indicate that the *father—pater familiae*—becomes a prisoner as a result of his power and authority at the home. At a closer look, hostility is exerted against the guest for the host to regulate his temporal conditions of inferiority. When this happens the guest is transformed into an undesired guest (Derrida names as *a parasite*). Following this, Daniel Innerarity (2017) calls attention to the harmful effects of fear, such as global risks, which close the borders of nations subordinating and effacing aliens' presence. He draws a parallel between risks and guests. Both ask for hospitality and appear suddenly to interrogate society. The current obsession for zero-risk society as well as the anti-migration bans corresponds with an epochal moment when hospitality is dying and Western culture withering away. Zygmunt Bauman has widely explored the dichotomies of double-structure mobility where few new persons are legally encouraged to travel worldwide whereas the rest of the mankind is subject to economic exploitation and immobility—a new emerging world of vagabonds and tourists, where the privileged position is determined by the capacities to impose surveillance technologies oriented toward detecting and punishing illegal mobilities (Bauman, 2017). Though criticized, Bauman's insight ignited a hot debate revolving around the power of tourism as an agent of postmodernism and neocolonialism. Per this academic standpoint, Bauman's classificatory schema obeys a dichotomy between voluntarily and involuntarily travel which very well escapes to empirical observations in the Global South, at best in the ways and terms he has formulated them (Marotta, 2002; Jacobsen & Marshman, 2008; Franklin, 2003; Welten, 2015; Ray, 2021). Finally, Korstanje's thesis on the death of hospitality recurs to 9/11 and the rise of global terrorism as an external force that tends to erode the sacred laws of hospitality, which is one of the mainstream cultural values of the Occident. Whereas the European powers colonized the world in the eighteenth century by introducing curiosity and science, terrorism has accelerated a crisis of sense when the "non-Western Other" is suspected as an enemy living within (Korstanje, 2017). Is COVID-19 a continuation of the same tendency where the "Other" is negated?

Conspiracies and Literature

David Kelman (2012) is a well-renowned expert who has explored the sociological nature of counterfeit politics and conspiracy theory. In his book, *Counterfeit Politics: Secret Plots and Conspiracy Narratives in the Americas*, he argues convincingly that classical scholars in the fields of political science adopted a negative connotation about counterfeit politics. They punctuated this form of politics followed a pathological nature, which is caustically organized to instill fear and paranoia. Unlike the discredit this genre implies, Kelman theorizes on conspiracy theory as the ideological core of modern politics, present in all Western cultures. Crisscrossing from Latin America to Anglo-America, the genre marks a gap between the official and unofficial stories. Conspiracies, so to speak, broadly revolve around problems or undefined situations that may jeopardize the system. In this way, hapless citizens are not familiar with their future or what political actors plan secretly. The secrecy derives from the urgency of these actors to design policies that otherwise can very well be rejected. As Kelman eloquently noted, these machinations exhibit temporal efforts to unravel what is hidden in politics. Western democracies historically evolved according to a conspirational fable to legitimize their authority in the threshold of time. For Kelman, literature offers a fertile ground to understand better the nature and evolution of conspiracy plots. His thesis holds that politics do not center on ideology as a dominant element of power, but it is somewhat constrained into a conspiracy narrative that separates the official truth from the secret. The legitimacy of politics consists of creating two bands, them and us, to construct a tensional relationship, antagonist relation. While a visible story (official discourse) monopolizes knowledge, a double structure where this knowledge is disputed emerges. Kelman suggests that conspiracy is not a symptom of political crisis but a precondition for politics. Social scientists are accustomed to thinking of politics as a stable entity but it cannot exist without the creation of an enemy, or a scapegoat, who helps in organizing the proper troop. The contrasting voice invests "a state of emergency" where the "Other" threatens the archetypes of an "idealized society". The figure of good politics plays a leading role in the configuration of modern detective novels. Per his viewpoint, the conspiracy represents a gap between political imagination and social reality that the villain fills. Having said this, politics is paradoxically legitimated by the same anxiety and mistrust it generates. Conspiracy, far from altering the status quo, affirms the legal institutions or the official discourse, it at the bottom denies.

A chapter that discusses the role of conspiracy theories in literature ultimately cannot be ended without touching, if not debating, Luc Boltanski's work, *Mysteries & Conspiracies*, which was recently translated into English and published by Polity Press. With an introductory foreword by Mark Greif, the book centers on counterfeit politics, conspiracy narratives, and the sense of reality. As Boltanski eloquently explains, if psychiatry gradually adopted and developed the phenomenon of paranoia as a set of symptoms oriented toward replicating a stable state of delirium, sociology did the same with the conspiracy discourse. Counterfeit politics has been historically considered a pathological form of politics by classical literature, but for Boltanski, the problem should be traced back to the nineteenth century, a period which coincided with the urgency of reinterpreting reality through a new lens. A coherent explanation of conspiracy associates with the problem of power and how it is fabricated, circulated, and held. To some extent, an insight into the nature of conspiracies entails a radical interrogation placed on the world-making process. Boltanski argues convincingly that everything that occurs in the external world can be internally interpreted through what he dubs as "reality". Any mystery comes from an event whose character seems to be abnormal.

> A mystery arises from an event, however unimportant it may seem, that stands out in some way against a background … or against the traces of past event, not witnessed by the narrator, that remains perceptible later on. This background is thus constituted by ordinary understandings as we know them through the intermediaries of authorities (educational in particular) and/or through experience; the latter gives actions a relatively predictable character, especially by associating them with habits. (Boltanski, 2014: 3)

In consonance with Kelman, Boltanski says overtly that reality is divided into two different narratives—REALITY and Reality. While the official story narrates events following the status quo carefully, the unofficial story, though it cannot be fully validated, introduces suspicion to make the world more coherent. Having said this, Boltanski distinguishes the fantastic stories from the picaresque texts. While the former refers to a strange world that should be domesticated, the latter signals its incoherence. In this way, strangeness and incoherence rest on an enigma that shakes the social order. Once the enigma is finally resolved, order is restored. To understand this point, we need to step back to the notion of reality in Boltanski. It can be traced back to the seminal text entitled "The fragility of reality", which critically discusses the need to adopt pragmatic sociology to re-found the epistemological tenets of the discipline. For some reason, which is hard to explain here, classic sociology went through a clear and much deeper dogmatism precisely after the 1970s; although Bourdieu shed some light on the problem of social reproduction, he left some questions open to date. Bourdieu failed in combining the micro with the macro-social landscape, giving much attention to the status quo over other variables. The conspiracy theory was systematically overlooked in his sociological debates. The habitus constitutes a disciplinary approach originally forged by the ruling elite, controlling, if not effacing, the notion of chaos and uncertainty. As Boltanski observes, it is necessary to return to the earlier works of Goffman to re-signify "the structure of the situation" which follows an assortment of reasons. In Boltanski, to cut a long story short, reality exists through the social agency's lens, which interprets and adapts it assigning a proper understanding of the external world. In other terms, there are no facts but only events which are re-interpreted internally by the self that is biographically conditioned. Sociologists often play a similar role as linguists; they illuminate in a certain field, changing and reconstructing the social world they observe at the same time. This suggests that knowledge production and experiences are inextricably intertwined (Boltanski, 2014).

Boltanski goes further when he highlights that the *constellations of reality* are grounded in a set of durable uncertainties, which range from the law to the social customs. The conspiracy and detective novels go against "the background", which surely means "against everything real" in the world. Through these regularities, society constitutes revolving around a historical context. Not surprisingly, the advances of science and technol ogy, in the eighteenth century, opened the doors to many uncertainness and anxieties which were very well rechanneled through conspiracies and detective novels. In terms of Boltanski, the notion of reality, as it was imagined by Enlightenment, faces gradually a crisis where the official discourse dissociates from a new unofficial reading of events. What social, detective, and spy novels have in common is "the conspiracy form", which can be defined as the attempt (inquiry) to see an object beyond society's social relations.

> This deliberate, functional layering of identities characterizes the structure of what we may call the conspiracy form. A conspiracy is an object that is only perceived as such—as distinguished from ordinary human relations—from the outside. It is distinguished from ordinary relations only by the operations of unveiling that sets an apparent but fictitious reality and a hidden but real reality side by side, on the same level. (Boltanski, 2014: 13)

Boltanski coins the term "the plain of reality" to exhibit each agent's function (role) in the system. For example, if I take a taxi I assume the driver will intend to bring me to the mentioned address. When this does not happen, suspicion arises. To put it differently, conspiracy denotes a previous inquiry that "changes and interrupts" the normal course of action I am referring to. It leads the inquiry to unexpected situations which were never demarcated or imagined. Boltanski's main theory holds polemically that paranoia and detective novels' psychiatric symptoms are two sides of the same coin. Originally depicted as a mental illness, paranoia extends the lens of inquiry toward an inexpugnable place, a dark point where reasoning does not reach. The investigator, in police novels, serves as a person with paranoia, though healthy. The eighteenth century witnessed a dual context; on the one hand, technology and science made of daily life a more predictable environment, carefully organized and predictable according to capitalist means of production, while, on the other hand, Western societies went through states of anxieties and fears that evinced a clear fragility. Detective novels proved work to be alleviating instruments that kept society united. In a nutshell, the rise of detective fiction dramatized what Boltanski calls "the anxiety derived from the reality of reality".

Doubtless, the logic of conspiracy is given by the architecture of the nation-state, if not the fear of the stranger. The modern state was legitimated by articulating disciplinary mechanisms (à la Foucault) such as statistics and the sentiment of nationhood but operated on the lines of a rational instrumentalization where different ethnicities were forced to live together within the limits of the territory. The notion of reality gives citizens stable parameters of existence that are based on the common law, traditions, and democratic institutions. The detective represents an agent of state whose "primordial inquiry" restores the lost order to an abnormal situation. Hence, the detective, the state, and the novels seem to be subjected to what Boltanski calls a "trial", or a test, where the transgression is finally corrected. What is more important, the state's legitimacy is not determined by its ability to placate anxieties but rather by the urgency of exacerbating (harnessing) the natural fears and anxieties that are typical of politics. This happens because:

> The trial that faces the State in detective fiction is precisely the mystery as an anomaly concerning reality. The reality, which the State is supposed to underwrite, is not called into question by the fact that there are criminals, a fact that in a way constitutes the very justification of political order, but by uncertainty surrounding the circumstances of the crime, and especially its attribution, since any character at all, however irreproachable he or she appears, may be its author. Thus detective fiction dramatizes a reality open to suspicion in all respects, material as well as human, physical as well as social. (p. 19)

What are these anxieties or at least what shapes do they finally take? As Boltanski notably remarked, the first anxiety is related to the need to reach the truth or for investigating the truth to monopolize the speech. The nation-state looks to know what happens and, paradoxically in doing so, open the doors to an unofficial version of facts. The dissociation of official and unofficial stories embodies a puzzling situation the state is culturally embraced to resolve. The second anxiety consists of the wide extension of suspicions. The crime, like the figure of risk, echoing Beck, places all citizens in equal conditions before the investigation. No matter the social class, the position or status in society, or the social hierarchy, all are potential criminals in the detective stories. Finally, the relation between the state and the asymmetries of capitalism occupies a central place in understanding the priviledged elite conspire against the country. In these types of stories, capitalism's social inequalities are brought to the foreground, as well as the vices in powerful and wealthy people. The organization of the nation-state, in any way, is centered on multiple dynamics and contradictions. To wit, Boltanski uses the term *territory and flows*. The former describes everything real and certain in our lives, while the latter interrogates the established ruling class, probably marking the material inter-class inequalities. Having said this, he notably acknowledges that democratic governments, at least those having parliamentary representation, are prone to stimulate conspiracies and detective stories than authoritarian regimes where the sense of reality is one-sidedly fabricated and, of course, unilaterally imposed. If the hero embodies all that is good in a grim world, he is probably an outsider, an outlander, or, what is worse, an outlawed person who struggles to shed light on the dark mysteries. On the opposite, the villain has the great intelligence to change the borders of reality and pseudo-reality, but he is a person of power, status, or someone who belongs to the "privileged group". As Boltanski insists, sociology has some commonalities with conspiracy discourses and detective stories. It departs from the urgency to decipher the "reality of the reality" questioning established dogmas or putting them under the critical lens of scrutiny. Conspiracies offer a spectacle to the readership, a type of game where two tensional forces converge. On the one hand, the formal institutions were originally designed to protect lay-citizens while safeguarding the social order. On the other, the corrupt police or politicians loom secretly to impose "a brute force" without any ethical concern or moral justification. This begs a puzzling question: what is the meaning of the idea of going against the background that Boltanski obsessively mentions?

To put things simply, detective novels (e.g., Sherlock Holmes stories) describe a complex and mobile (Victorian) world that was culturally cemented through imperial expansion. The different actors, moving here, come from a privileged aristocratic class that benefited from imperial rules and the rise of a new rich class, which is considered brute and uneducated. The servants, of course, are portrayed as irrational children who should be constantly monitored and whose education should be oriented to be patronized by their masters. In a climate of great asymmetries, the norms and rules invariably take a stable condition. Nonetheless, as Boltanski contends, this state of normality radically changes when the status quo feels jeopardized. At a closer look, the state seems to be authorized to meddle with public life, when circumstances oblige, pushing legality to a peripheral position.

> It is against a background of normality that the singularities stand out, in such a way that the subjects whose uneasiness they arouse manifest what can be called their ordinary sense of normality. This sense is implicitly understood as a form of common sense, which is naturally also present in the detective. Through the intermediary of this common sense of normality, the uneasiness of ordinary characters who are blinkered by the limits of their functions can arouse the interest of the detective and stir up his boundless attitude for reconstituting all linkages through which reality holds together. This is where his exceptional intelligence resides. (Boltanski, 2014: 49)

The above-noted extract shows two important things which need to be discussed. On the one hand, the detective takes part in a first-order class educated to act with executive power. Unlike servants, in particular, imperial servants who are indoctrinated only to serve their masters, the detective penetrates the dominant discourses to illuminate the readership or unveil the reality of the reality. On the other, the detective is an empowered agent who protects the law but, paradoxically, he moves beyond the law. All are equally suspected before the law, as Boltanski suggests, in the realm of conspiracy plots. At this point, legality and reality coalesce in a hybrid form. With the benefit of hindsight, it is possible to distinguish the detective from the ordinary policeman. While the policeman is limited to seeing the world in the borders of legality, the detective can see the evil in its pure form. It is important to add that the detective moves through the articulation of two types of violence. The (symbolic) verbal violence, which interrogates the reality critically as it is taken for granted, explains his success when the mystery is resolved, and the physical violence is often employed to reduce the villain's resistance. Without going into further details, the policeman is, above all, an officer, so to speak, an agent of the state who obeys, under all circumstances, the mandate of the law, even when this law is corrupted or tainted by human greed. However, the "private" detective works as an anti-hero who does everything that should be done, sometimes violating the law. The detective is part of the private life but moves freely in the public one as well. This reasoning exhibits a clear-cut dualism between the structure of power and the agency's constitution. This dualism leads Boltanski to question why conspiracy theories and detective stories entertain us generation after generation.

Secrecy and Detective Stories in Review

Detective stories rest on the urgency of sublimating a finely enrooted dualism that is typical of capitalism. A quintessential contradiction mainly marked the passage from the despotic feudal state to the liberal state. The feudal lords were immune to the action of the law, while peasants were subjected daily to countless risky situations and unjust accusations. By placing all men in equal conditions under the law, a mandate introduced by liberal thinking, the state put all men equally suspect before justice. In the eighteenth century, the invention of alterity allowed citizens to reach a "serial totality" where the neighbor becomes unknown. To understand the reality, citizens alluded to "interlocutors" who regularly described the reality from the eyes of the state. This opened the doors to a dual identity where the self and the "Other" coexisted. As Boltanski reminds, this is exactly the plot of Dr. Jekyll and Mr. Hyde. The plot speaks to us of a respectable doctor who is not subject to any suspicion, who melts into an unknown personality, a monster, or a criminal. The modern nation-state devotes efforts to regulate daily life through the articulation of the dominant structure. Still, in doing so, it confers the possibilities to move freely within the borders of nationhood. In regard to the newly emerging capitalist societies, Boltanski goes on to write:

> These societies are profoundly ambiguous in the sense that the identity of persons is defined with reference to contradictory processes. On one hand, we find processes that assign persons a defined position with the structure of domination, among sexes, races, nations, orders, and—above all, with the triumph of capitalism—social classes; on the other hand, we find processes of emancipation that offer the same persons the ideal of formal equality on a purely legal basis. (p. 67)

Having stated this, the genre took a different direction when it arrived in France. Unlike British novels, where the private detective plays a leading role in resolving mysteries, the figure of the detective seems not to be relevant in French novels. The commissioner occupies the position of the detective, but he never interrogates in the public sphere. The French plots describe situations of greed and conflicts for money, all of them placed in the private sphere, instead of conspiracies, if not attempts, to destabilize the established order (like in Anglo-Saxon literature). Apart from this, similar elements such as dualism between the law and legalism and practical reasoning or the inconsistencies of inter-class inequalities, revolving around mysteries, are common in both kinds of literature. A question immediately arises: what is the relation between terrorism, COVID-19, and conspiracy genre?

As Boltanski indicated, detective stories originally surfaced as a direct consequence or, more appropriately, a clear sign of the triumph of capitalism over other forms of production. The West has expanded through the introduction of alterity, a figure who populated in novels, literature, and other forms of cultural entertainment during the colonial rule. The conception of hospitality was essentially invoked to legitimate the European expansion, as well as the so-called inferiority of native tribes in the Americas. As already discussed in earlier chapters, it is important to trace back Anthony Pagden's (1998) outcomes which reveal that the conquest of the Americas was finally legitimated because some aboriginal tribes were not familiar with the rites of hospitality and the right of free transit. For Spain's philosophical circles, since the idea of free transit was culturally enrooted in the universal rights, those who were not cognizant of these rights had to be treated as sub-humans, as Pagden eloquently remarks. In a nutshell, hospitality and colonialism were inextricably intertwined. By the same token, the "non-Western Other" was of paramount importance to keep the European readership under control. Without mentioning the literature of the alterity, the spectacle played a leading role in affirming European ethnocentrism and its cultural values. The 9/11 attacks subverted the relation between selfhood and the Otherness as never before. This event alimented the belief that the enemy does not come from the outside, but he lives and behaves like us, even looking like a lay-citizen. The detective stories emulate the dichotomies between ideology and reality. The "Other" was not an object of curiosity, which reaffirms the European superiority, but also a radical enemy (a villain, a monster) who should be systematically eradicated. The war on terror, following this, gradually closed the Western economies to the "undesired Other", affecting seriously one of the cultural values of capitalism: the sacred laws of hospitality.

Lastly, COVID-19 continues the same tradition, not only locking the social ties but also undermining the trust. Today's experts warn about a new war, a war against "an invisible enemy". If the message of terrorism alerted on the possibilities that anyone at any time can be a potential terrorist, the days of COVID-19 speaks of the possibility that all of us would be potential carriers (terrorists) who transmit the virus. In a classic world, the "alterity" is an object of fear and curiosity, but now it can be radically negated to "an undesired guest". This is a radical world where tourists and strangers are not welcomed any longer, and new geography marked a feudalization process characterized by chauvinism, racism, and supremacist discourses. Having said this, conspiracies offer an efficient form of spectacle and mark the current politics. Conspiracy is the other side of ideology, to put things simply. To illustrate this better, let me return to the problem of terrorism and 9/11. The days after 9/11, the media transmitted the idea that the US was the only superpower whose hegemony expanded worldwide, just after the collapse of the Soviet Union. This American superiority, from the outset, oriented toward exporting the ideals of equality, freedom, democracy, and free trade to undemocratic societies with the end of strengthening the collective well-being of citizens. But this has a clear contradiction. If the US is the benevolent superpower with an all-seeing eye, why has 9/11 occurred? Or at least why was this attack not prevented?

A tentative conspirational answer alludes to the fact that authorities needed an excuse to take control of strategically important countries in the Middle East. In doing so, authorities not only were aware of the imminent terrorist attack planned by Al-Qaeda but they temporarily turned a blind eye to the New York airspace. In sum, the US promoted and allowed the terrorist attack to conduct a process of retaliation in the oil-producer countries. As Kelman and Boltanski suggest, conspiracy theories are based on a conceptual gap that is filled with speculations and abstractions that are never verified. Conspiracy represents the other pole of ideology filling the missing piece to accommodate the ideological architecture. Sociologically speaking, conspiracy exhibits an effort to make controllable what in essence is not. Whenever reality overrides fiction, conspiracy responds to the questions ideology keeps open. Accusations of conspiracy theories work as a mechanism to validate the official discourse, silencing those voices, bringing a radical shift to the system. Conspiracy theory functions as an epithet which plays an important part in ideology, not social or political theory. Last but not least, in psychology, conspiracy and paranoia are not pathologies; rather, they are adaptive mechanisms of the frustrated ego to an intransigent environment. Conspiracy theories surface to explain the failures of the ego to reach life goals, and they can lead to the rejection of the "reality principle". The same applies to the conspiracy rumors indicating China was responsible for the COVID-19 outbreak.

Conclusion

The quintessential feature of conspiracy and counterfeit politics, without any doubt, associates to the nation-state or at least what Boltanski dubbed as "negative anthropology", that is, the mere possibility that all actors are equally suspected of being potential murderers. This seems to be the logic that conspiracy, terrorism, and COVID-19 ultimately share. Citing Hobbes (1998), Boltanski returns with the idea that lay-citizens confer the legitimacy of violence against a third party with the end of avoiding the "War of all against all". Following this, the state not only concentrates efforts to make the public life stable, alerting on the risks of returning to a wild state of nature, but keeps the monopoly of the law, which is a communal understanding of what can be done or not. The conspiracy plot subverts the axiom that says "everybody is innocent until proved guilty".

Like the post-9/11 context, COVID-19 seems to enhance a much deeper understanding of the conspiracy plot, which is finely enrooted in Western civilization's cultural background. To be more exact, the upsurge of this literary genre and fear should be explained by the need to legitimate the disciplinary apparatuses of the state. Like the medical gaze, which is omnipresent to intervene in the body when it is sick, even extirpating an organ when life is compromised, the detective emulates the all-seeing eye that dissects into the reality of the reality. What is equally important, the nation-state instrumentalizes the disciplinary mechanism of violence through an invisible enemy's invention. The same can be said regarding terrorism and the ghettoization process. Like a virus, the terrorist looms in the secrecy of clandestine life, appearing to be a normal citizen, but always ready to attack. He looks like us and is educated in Western universities but what is logically more important he hates the society he belongs to. Emulating the medical reasoning, counter-terrorism policies allow security forces, like doctors, to lockdown suspected actors, even placing entire (Muslim) neighborhoods under siege. Echoing the metaphor of cancer, which urges to be extirpated before future ramifications to the entire body, terrorism leads the West to close the borders to "non-Western Others" and locate rapidly and eradicate consequently potential (infected) cells to avoid political instability. As Korstanje (2017) puts it, hospitality is dying because a recent radicalized discourse in the Occident negates the alterity and because the Muslim community is running a serious risk of being demonized, sub-humanized, and subsequently extirpated.

References

Allen, C. (2010). *Islamophobia*. Ashgate Publishing, Ltd.

Altheide, D. (2017). *Terrorism and the politics of fear*. Rowman & Littlefield.

Bauman, Z. (2017). Tourists and vagabonds: Or, living in postmodern times. In *Identity and social change* (pp. 19–32). Routledge.

Bauman, Z., & Lyon, D. (2013). *Liquid surveillance: A conversation*. John Wiley & Sons.

Boltanski, L. (2014). *Mysteries and conspiracies*. Polity Press.

Boltanski, L., Rennes, J., & Susen, S. 2014. The fragility of reality: Luc Boltanski in conversation with Juliette Rennes and Simon Susen. In *The spirit of Luc Boltanski. Essays on the pragmatic sociology of critique* (pp. 591–610). Anthem Press.

Derrida, J., & Dufourmantelle, A. (2000). *Of hospitality*. Stanford University Press.

Franklin, A. (2003). The tourist syndrome: An interview with Zygmunt Bauman. *Tourist Studies, 3*(2), 205–217.
Hobbes, T. (1894). *Leviathan*. Routledge.
Hobbes, T. (1998). *Hobbes: On the citizen*. Cambridge University Press.
Howie, L. (2012). Witnessing terrorism. In *Witnesses to terror* (pp. 155–175). Palgrave Macmillan.
Innerarity, D. (2017). *Ethics of hospitality*. Taylor & Francis.
Jacobsen, M. H., & Marshman, S. (2008). Bauman's metaphors: The poetic imagination in sociology. *Current Sociology, 56*(5), 798–818.
Kelman, D. (2012). *Counterfeit politics, secret plots and conspiracy narratives in the Americas*. Bucknell University Press.
Korstanje, M. (2017). *Terrorism, tourism and the end of hospitality in the West*. Palgrave Macmillan.
Korstanje, M. (2019). *Terrorism, technology and apocalyptic futures*. Palgrave Macmillan.
Laster Pirtle, W. N. 2020. Racial capitalism: A fundamental cause of novel coronavirus (COVID-19) pandemic inequities in the United States. *Health Education & Behavior*, 1090198120922942.
Marotta, V. (2002). Zygmunt Bauman: Order, strangerhood and freedom. *Thesis Eleven, 70*(1), 36–54.
Milano, C., Novelli, M., & Cheer, J. M. (2019). *Overtourism and tourismphobia: A journey through four decades of tourism development, planning and local concerns*. CABI..
Pagden, A. (1998). *Lords of all the world: Ideologies of empire in Spain, Britain and France: c1500–c1800*. Yale University Press.
Ray, A. 2021. The vagabond as nemesis of the tourist: Towards a postcolonial critique of Zygmunt Bauman. Special Issue 'Critical thinking in tourism'. R. Tzanelli & M. Korstanje (Eds.), *Tourism, Culture and Communication*, 21(2–3): 107–116.
Ritzer, G. 2019. Inhospitable hospitality. In B. Rowson & C. Lashley (Eds.), *Experiencing hospitality* (pp. 73–91). Nova Science.
Selwyn, T. (2019). Hostility and hospitality: Connecting Brexit, Grenfell and Windrush. In B. Rowson & C. Lashley (Eds.), *Experiencing hospitality* (pp. 51–72). Nova Science.
Skoll, G. R. (2016). *Globalization of American fear culture: The empire in the twenty-first century*. Palgrave Macmillan.
Springer, S. (2020). Caring geographies: The COVID-19 interregnum and a return to mutual aid. *Dialogues in Human Geography*, 2043820620931277.
Welten, R. (2015). Hospitality and its ambivalences. On Zygmunt Bauman. *Hospitality & Society, 5*(1), 7–21.
Zizek, S. (2020). *PANDEMIC!: Covid-19 shakes the world*. John Wiley & Sons.

CHAPTER 8

Let Me in! The Affirmation of a Radical "Other"

Introduction

As P. Pearce (2019) puts it, *travel behavior*, far from being a static object of study, varies depending on time and culture. People are psychologically motivated to visit different sites while having countless experiences. The chapter reflects with empirical information and describes in what direction this travel behavior is going. The first section synthesizes a short introduction of the archetype of the vampire, a fictional actor who corrupts the essence of hospitality. The vampire is an undead "Other" who attacks his victims, suctioning all the blood, but he should be invited to enter in. The guest voluntarily opens the doors to this inenarrable evil. The same can be said of COVID-19 and a new war, which occupies a central position in the second section. The third section offers a complete empirical basis that helps readers expand the current understanding of the industry's changes in the post-COVID-19 days. If the nation-state was a political unity in the past centuries, emanating a centralized authority, it is now being decomposed into a decentralized and atomized sub-unit. The case duty discussed based on Argentina but applicable to other countries in the world probes two important things.

In the context of an emergency, the doctrine of free transit and circulation is automatically suspended. This leads each province (sub-unit) to adopt self-efficient migratory restrictions while confronting with the right

M. E. Korstanje, B. George, *Mobility and Globalization in the Aftermath of COVID-19*,
https://doi.org/10.1007/978-3-030-78845-2_8

to travel. The articulation of various travel bans accompanies the atomization of the nation-state. In our study, new travel requirements which are often requested for international travels, such as health cards, medical assistance coverage, and medical travel insurance, are necessary for domestic journeys. What is equally important is that showing an immunity passport to travel is essential in understanding this new era. A passport is defined as a document issued by each state certifying the lack of any criminal record to be safely accepted by the hosting state. The passport successfully validates that we are who we say to be. The immunity passport was a project thought to issue a type of document that certifies the immunity of citizens, probably because they were innoculated or infected by the COVID-19. People were supposedly allowed to move because they obtained a natural immunity against the virus. Similar to quarantine, the registration of infected people was a political attempt of states to control the pandemic. The issue will be discussed further in the concluding chapter.

A Short Debate: COVID-19 and Tourism

By the end of December 2019, a new virus outbreak originated in Wuhan, China, and paralyzed not only the tourism industry but also the global trade. SARSCOv2, known as COVID-19, has invariably led the industry to an unparalleled crisis (Radic et al., 2020; Prayag, 2020). In a matter of months the virus propagated to five continents situating a global pandemic as well as one of the nightmares of Western civilization. Because of this, scholars moved to analyze the future of tourism and debated different steps and protocols to mitigate the negative effects of the pandemic (Rogerson & Rogerson, 2020). The pandemic altered society's means of production, which means the macro-economic forces of production and consumption and the internal protocols in tourist establishments to maximize the tourist and front desk staff security (Filimonau et al., 2020). Some voices call attention to the employment of robots to minimize contagion risks (Seyitoğlu & Ivanov, 2020).

In the incipient literature, theories revolving around COVID-19 and tourism are mainly organized into two clear-cut families. On the one hand are those studies interested in developing new methodological lines to overcome the current limitations of tourist-centricity (Ivanova et al., 2020; Wen et al., 2020; Altejevic, 2020), which is the obsession for interviewing tourists as the only reliable actors of the system. On the other are those studies that center on the opportunities of social distancing and degrowth to reverse the current levels of contamination (Gössling et al., 2020; Higgins-Desbiolles, 2020; Polyzos et al., 2020). This happens because the tourism industry is about to experience a new rebirth. In this respect, Ioannides and Gyimothy (2020) assert that COVID-19 gives mankind a new opportunity to educate leaders in the values of more sustainable tourism. The negative aftermaths of climate change can be reversed if leaders adopt more eco-friendly measures. It is important to mention that though another pandemic has hurt the tourism industry, only COVID-19, as a foundational event, rattled the basis of the system, igniting a never-seen crisis. Despite their differences both academic groups agree that COVID-19 is not only a disrupting but a foundational event that has radically changed the Western lifestyle and modern travel behavior. Although the incipient literature sheds light on the effects of COVID-19 in the industry the published works are speculative works that lack an empirically firm basis. Tourism-centricity, which means the obsession to take the figure of the tourist as the only valid source of information, in applied research has invariably left tourism researchers in a puzzle (Franklin, 2007). Here an interesting question arises: how can we study tourism in a world without tourists?

Let Me in

Over the years, the economic-based paradigm over-valorized the tourist as the main actor of the system while relegating other voices and agents to a peripheral situation (Franklin, 2007). Today, tourism research has a fertile opportunity to re found its methodological and epistemological bases toward a maturate option (Avraham, 2020; Hao et al., 2020). In earlier works, we recognized that far from being a foundational event, the coronavirus disease affirms a previous trend that originated in 9/11 and the war on terror originally declared by the Bush administration. By naming this epoch as "the death of hospitality" we certainly found some commonalities between COVID-19 and 9/11 (beyond the strange repetition of the numbers nine and one) (Korstanje, 2020). To some extent, terrorism ignited a new face of Western lifestyle where "the Other" was the subject of fear and mistrust. If the colonial travels were inspired by the need of meeting a radically different "non-Western Other", the culture of fear, which was shaped by terrorism, depicts the neighbor, who looks like us, as a potential terrorist (Howie, 2012; Rives-East, 2019; Korstanje, 2017, 2019). In the days of COVID-19, all citizens are potential terrorists (carriers) who threaten the social order. The recent video game *Among Us*, which captivated young gamers, serves as a clear example punctuating the idea of living with the enemy inside. This multiplayer game, which the American Studio Innersloth developed, takes place in a space-themed scenario, and players select the role of crewmates or of impostors who will sabotage the mission. Crewmates are motivated to identify and eliminate rapidly the impostors who try to sabotage and kill the crewmates. In this game, nobody knows the intentions and the role of the "Other"; suspected impostors can be eliminated through a plurality vote, which may be initiated by calling an emergency meeting. Crewmates ultimately win when all impostors are killed, and likewise, impostors (like terrorists) look to assassinate all crewmembers.

Secondly, and most importantly, digital surveillance technologies structured new forms of control at borders and airports, which further interrogate travelers. As Derrida and Dufourmantelle (2000) brilliantly observes, hospitality can be offered or not, but whatever the case may be, two questions are vital when we meet a stranger: who are you and what do you want? The figure of visa occupies a central position in the ritual of hospitality. The word *visa* stems from *visum*, a Latin verbal past tense associated with *videre*, which means "to see". The hosting state, echoing our French philosopher, needs to check our identity before proceeding to offer hospitality. To the traditional documents requested of travelers such as a round-trip ticket and a visa or passport, we must add the COVID-19 test and temperature test. Metaphorically speaking, in the war on terrorism technology scrutinized the travelers' luggage suspecting a bomb inside; now the threat is supposed to be a carrier by the traveling guest. The war on terror sets the pace for the war against a virus. The question of whether technocracy reminds the ideological system of governance based on technology to frame the decision-making process, "tourist travels" are shifting from *leisure travels* to *technocratic travels*. In post-pandemic tourism, the "Other" has been transformed into an undesired guest.

Thirdly, both terrorism and COVID-19 operate in the context of high uncertainty, which is characterized by the rise of "a feudalization process". In some words, let's describe to readers the situation in Argentina, and of course the world, these days. In March 2020, President Alberto Fernandez implemented countless measures to stop the contagion, which included the closure of airspace, borders, restaurants, and factories and the cancellation of public spectacles, events, and commercial activities (to name a few). In this respect, thousands of Argentinean tourists were suddenly stranded abroad. The foreign ministry played a leading role in the coordination of repatriations. At the same time, many foreign tourists wandered on the streets of Buenos Aires looking for a way to be repatriated to their homes. Something had to be changed. In the following days, police and migration officers patrolled the streets looking for foreign tourists who had violated the lockdown. They were systematically traced, put on trial, and exiled. All visitors from neighboring countries such as Brazil or Chile were required to be isolated for 14 days. The circulating narratives revolved around the tourists, who had been historically seen as ambassadors of civilization but had now been re-typified as an ambassador of disaster. Some assumptions reveal today that almost 4000 people were stranded between the provinces of Chaco and Formosa in November 2020. The lack of standardized methods to regulate mobilities in Argentina led the provinces to adopt different measures. John Urry (2002) eloquently imagined the figure of the tourist gaze to denote a much deeper cultural matrix which is finely enrooted in global capitalism, but now this tourist gaze mutated to a wicked gaze. The "Other" who is seen as a potential danger should be strictly monitored and controlled, if not eradicated, to preserve order. It is safe to say that COVID-19 seems not to be the cause but a symptom of a moral decomposition, a feudalized world where the "Other" is negated. We are thinking of the metaphor of the vampire who needs an invitation to step in. As Brian Aldiss puts it, vampires represent the embodiment of an ancient fear centered on the dead for the contemporary culture. We often fear the spirit of the dead, and of course, the figure of the vampire coincides with such a vision; we are compelled to dwell in the same household with the dead, which frightens us (Aldiss, 1997). Behind the philosophical dilemma of the vampire lies the problem of the "Otherness", as Joan Gordon and Veronica Hollinger (Gordon & Hollinger, 1997: 3) explain.

> The figure of the vampire, as a metaphor, can tell us about sexuality, of course, and about power; it can also inscribe more specific contemporary concerns, such a relation of power and alienation, attitudes toward illness, and the definition of evilness at the end of an unprecedently secular century.

In sharp opposition to humanity, monsters help to reconstruct our contemporary cosmologies. Of course, these cosmologies change in the threshold of time. Each epoch has its own understanding and representation of the monstrosity. Bram Stoker's machinations of Dracula differ notably from the present horror films and novel plots where the vampire is more sympathetic of human vulnerability. Still further, the vampire epitomizes the metaphor of mass consumption, which leads toward a never-ending hunger (Gordon & Hollinger, 1997). By the same token, Jules Zanger (1997) distinguishes between the old and the new vampires. While the former adopts a lonely life isolated from civilization, as Stoker's Dracula, the latter signals the need for forming a communal family, a community of "others" under the auspices of the blood. Examples of this are Anne Rice's Vampire Chronicles (1976) and The Lost Boys (1987). The old version of Dracula presents the figure of the vampire as a radical (pure) evil that should be eradicated. The modern version appeals to more humanist morphologies where the vampire struggles only to survive (jointly with others like him who are considered parasites or monsters). The modern vampire lacks the anti-Christian dimensions given by Stoker. In opposition to Dracula, the *good vampire* emerges not only as a new postmodern version of this evil but he would be a businessman, a journalist, a writer, or a rock star. He is not a serial killer, only a victim of his nature. Ultimately, Nina Auerbach (1995) argues convincingly that vampires condense the repressed fears of society taking different shapes in time. In Victorian times, McCarthyism, the Cold War, or even today, vampires take different bodies and significations. Most certainly, vampires are immortal and smarter and stronger than the man. Although they are entitled to live forever, changing over time, they symbolize not the fear of death but the fear to live. Their evolution as eternal species depends on the capacity of infecting (converting) others. In Auerbach's term*s*, "They are fearful reminders of the infinite things we are not" (1995: 6).

As previously argued, vampirism, like many other monsters, associates strictly with the human impossibility to give hospitality to "the stranger Others", a point which experts systematically overlooked. Let us step back on this concept for a moment of our inquiry—Dracula, the vampire, as the supreme evildoer, needs to be invited before he can drink the blood of his victim. The invitation occupies a central position in the radicalization of evilness. In Stoker's horror novel, the story begins with an English lawyer, Jonathan Harker, who visits Count Dracula in Transylvania, Moldavia. Dracula requests legal support for a real estate transaction. Harker soon realizes that he is a captive of Dracula and meets three vampires rescued by Dracula. Finally, Harker, Dracula's guest, is abandoned by Dracula, the master of these vampires, but he escapes . Meanwhile, Dracula embarks a ship called the *Demeter* and travels toward England, where he purchases various estates under the name of Count de Ville. Once in England, Dracula desires Lucy but he needs her invitation to step in. In a nutshell, Stoker's gothic novel symbolizes the intersection of hospitality, or at least the impossibility to offer hospitality with evilness, with pure evil. The invitation plays a crucial role in the configuration of hospitality. In the same way, the demon needs an invitation to possess his victim too. This suggests that the nature of evilness is given to those who violate or are unable to offer hospitality to their guest. Without any doubt, the lockdown accelerates some material asymmetries as well as the suspension of the right of traveling, at least through the territory (Korstanje & Olsen, 2011; Korstanje & Tarlow, 2012). Throughout the constellations of Western imaginary, the vampire (as an undesired Other) is comparable to COVID-19 or the terrorist. The virus takes its hosts disposing of their bodies to replicate itself, exhausting the vital resources to death. The terrorist, as we have seen in a former chapter, lodges in society similar to the *parasite* which kills its host. Like the virus or the vampire, the terrorist lives and educates in the Western lifestyle; he even looks like us. But what makes him a monster is the fact that he can suddenly develop any time radical hate against the hosting society. Such a discourse ushers in the West a climate of extreme anxiety and xenophobia.

In present feudal Argentina, provinces closed their borders and probably tightened the controls and migratory documents. The main contention of this short note is that we are living a feudalizing process vividly marked by a strong fear for the stranger. Scholars should pay attention to the sociological effects of COVID-19 on culture in the years to come. What is equally important, some provinces attune efforts to boost tourism to revitalize their economies, opening their internal frontiers. While visitors should present the respective COVID test document certifying they are healthy, other requirements such as an all-inclusive tour or round-trip tickets without mentioning certificate of residence or health insurance for travels are strictly requested upon arrival. All these noted measures are requisitions for traveling abroad but hitherto not applied for domestic tourism in normal conditions.

Last but not least, we have discussed how global geography is undergoing radical transformations. The question of hospitality is of paramount importance to understand not only the radicalized discourses of hate, against foreigners, and the social decomposition of the current epoch. Islamophobia, tourist-phobia, Windrush scandal, Brexit, hostilities against

Asian tourists who are suspected of disseminating the virus, the Wall in Mexico-US border, and now social distancing are clear symptoms of the end of hospitality, at least as we know it. The (non-Western) "Other" has become an undesired guest!

Thanaptopic Society

In some respects, terrorism and the attacks on the World Trade Center remind two important things. On the one hand, a radicalized group (Al-Qaeda) humiliated the most powerful country in the world employing four civilian airplanes which, so to speak, were valorized as the hallmark of modern mobilities. From that moment on, lay-citizens asked for tighter controls at the borders and kept away from public life. This gradual process of segregation has had durable effects on the psychology of the mind. Living at home, or stay at home, is considered the best option in an ever-changing world. On the other, the securitization process introduced after 9/11 speaks to us of a new climate of fear, characterized by Jean Baudrillard (1994) as "the great simulacra". In a futurist world organized to achieve a zero-risk society, technology governs our bodies from the future imposing hypothetical events which never take place as imagined. Home exhibits a fertile ground to re-connect inert bodies in the plug of a pseudo-reality. The curiosity for the Other's death and the obsession for consuming disasters, in a thanaptopic society, are signs of the same social decomposition we dubbed as "the end of hospitality" (Korstanje, 2016; Tzanelli, 2016, 2018). As Korstanje (2016) laments, the attacks on the US in 2001 not only changed the geopolitical relations among nations but introduced a new climate of mistrust and fear that transformed the urban social imaginary. If the modern sociology of Ulrich Beck or Lord Anthony Giddens focused on the position of risk as a mediator between citizens and their institution, 9/11 turned to a new stage of capitalism "where the Other's death" was the commodity to exchange. We passed from a risk society to a thanaptopic society where the spectacle of disaster delineates human relations. We dubbed this as *thana-capitalism*. From its outset, capitalism stimulated a Darwinist climate of competition where only the fittest players (if not the strongest ones) survive. The evolution of capitalism crystallized the obsession for gazing as an instrument to possess the "Other". Of course, in a culture based on ocularcentrism, nobody believes in what cannot be seen. This generated a decline of religiosity (the belief in the afterlife) and in the sacred law of hospitality which was an ancient tie between mankind and divinity. Strangers, in this way, should be protected any longer simply because men dreamed of the possibility of living forever. The possibility of the last journey, as discussed by French philosophy, foreclosed because of modern capitalism's medical and technological advances. In the twenty-first century, lay-people die less than in other centuries, but this death has become more savage. The terror toward death situates as the main psychological drama of Western civilization. This makes some sense because subjects do not believe in afterlife. The problem lies in the fact that the resistance to dying equates to a much deeper pathological behavior. Survivor syndrome is self-explanatory at this point. After a disaster, regardless of its nature, survivors often feel a much deeper sentiment of depression and trauma. To placate anxieties, some resilient mechanisms help survivors feel they are unique, touched by the Gods, because they have survived. This natural psychological mechanism of adaptation mutates to pathological conditions when it cannot be reversed. Survivors erroneously believe they are special—if not a superman, at least smarter and stronger than the rest. The spectacle of disaster often offers these types of narcissist characters. In thana-capitalism, citizens over-valorize their skills and possibilities to win, imagining that the winner takes everything. The narcissist character is deprived to work together with others to change the rules. Imagining that the Other's death implies a second chance to keep fighting for the top prize, the subject feels pleasure because he cleverly avoids death. Dark tourists are not only fascinated to consume the "Other's" death and pain, they valorize their institutions such as free trade, democracy, and capitalism as the best of possible worlds. The best metaphor that explains how thana-capitalism works is the plot of the novel *Hunger Games* (2012). In a dystopian future scenario, the nation of Panem is located in North America. The wealthy Capitol controls the surrounding twelve poor districts (colonies), where the conditions of life are hard. As a punishment for a past revolt against the Capitol, the Capitol holds a bloody game every year to which one boy and one girl from each district are invited. Each participant fights with the "Others" to death. Only one winner is possible and the winner takes all. President Snow legitimates his political authority promoting the logic of divide and rule. Participants who exaggerate their possibilities (skills) of becoming the winner are unable to cooperate with others. Katniss and Peeta, from District Twelve, change the situation when both realize cooperation is the only way of confronting Snow's authority. The same applies to the thana-capitalist society and the obsession (morbid gaze) for gazing at terrorism-related news. The 9/11 attacks have started a new process where the "Other" is commoditized, cannibalized, and neglected. One of the aspects that characterize the thana-capitalism seems to be associated with the fascination for the "Other's death". This point evinces not only in dark tourism consumption, a new drive or segment rapidly packaged and sold by marketing experts, but also in other forms of cultural consumptions as TV documentaries, novels, video games, and so forth. Based on a narcissist character, the self perpetuates its happiness through the "Other's pain". For the sake of clarity, the spectacle mounted around death ensures the pleasure maximization of the modern (thanaptopic) self. Needless to say, no other spectacle offers a *thanaptopic landscape* than the show of COVID-19.

Methodological Discussion

One of the limitations of applied research in a post-pandemic context is the impossibility of conducting face-to-face interviews. The lockdown forces us to employ digital platforms like Zoom, Gipsy, or Skype to bolster a fluid dialogue with the interviewees. We conducted 50 interviews with tour guides, tour agents, housekeeping staff, and front-desk staff in tourist firms. We followed the snowball method, which means that each interviewee recommends the next person to consult, creating a traceable and specific trajectory of interviews. The sample is not statistically representative and cannot be extrapolated to other universes. The interviews were recorded from April to the end of September 2020. The sample included 20 females and 30 males aged from 25 to 55 years and were geographically located in Buenos Aires, Mendoza, Cordoba, Salta, and Ushuaia. Given the impossibilities of conducting face-to-face interviews, the obtained results should complement future approaches. The nature of the investigation is an exploratory-descriptive one. The results are contrasted and tested against journalist articles and testimonies collected by secondary sources worldwide toward an all-encompassing model.

Argentina as the Main Case Study

Tourist Stranded

The desperate and failed attempts to contain the virus in Europe and the US led Argentinean president Fernandez to sanction a set of extreme measures which ranged from the immediate closure of borders and the airspace to a strict lockdown on March 20, 2020. This lockdown, known as *Aislamiento Social Preventivo y Obligatorio* [Social distancing precautionary and mandatory], was originally passed as a presidential decree for two weeks but extended for several months. The lockdown was composed of six different facets that involved different times and cities, each dotted with milder or stricter coercive dispositions. The presidential decree, which included the closure of borders and airports and the cancellation of all domestic and international flights and public and sports spectacles, led to serious losses for the tourism industry and the service sector. As a result, almost 40,000 Argentineans were suddenly stranded abroad. The foreign ministry gradually coordinated efforts to repatriate these conational citizens through the flights of Aerolineas Argentinas, Argentina's only flag carrier. Without any doubt, this was a moment of extreme anxiety and fear in the streets and airports throughout Argentina. Thousands of foreign tourists were trapped at airports waiting for their repatriation. Meanwhile, Argentinean tourists from the US, Europe, and Asia were brought home by the government. Those who lived in Buenos Aires were taken back to refurbished hotels to be kept in quarantine for 14 days (for prevention). This sanitary measure was suggested by the World Health Organization to avoid the rapid dissemination of the virus. The tourism industry ground into a halt given the suspension of international flights and the cancelation of arrivals of tourists coming from neighboring countries such as Chile and Brazil, the best segments of cities such as Bariloche or Lago Argentino. Of course, the negative impacts of the pandemic were not homogenous in all service sectors of the industry. While rent-a-car companies were marginally affected, about 30% of the hotel industry was on the brink of bankruptcy. In the same way, travel agencies, tour operators, restaurants, hotels, and cruises experienced losses in millions. The same apocalyptic scenario was experienced in the entire world. Stranded tourists felt that their governments simply abandoned them. Journalist Julia Logue reported the case of Edgardo Zeheiri and his family who were desperately trying to return to Argentina, similar to thousands of others in the same situation in Miami. The family was trapped after Fernandez declared the closure of borders and airspace (in mid-March). The couple was originally a budget to cost a short trip in which case the situation becomes from worse to worst. The situation was aggravated because of the high currency exchange between the dollar and Argentine pesos, besides the 30% taxes the Argentinean government imposed for purchasing in foreign currency.

Other tourists alluded to the lack of health care cover, including medicines or food delivery (Logue, 2020). In consonance with this, Alejandra Cansola (2020) reported that Mexico worked hard to repatriate thousands of Mexican tourists stranded in four continents. In addition to the human assistance these tourists needed, we have problems associated with health regulations and security issues each country demands. The official spokesperson of SRE (Mexico foreign affairs ministry) said:

> Mexico's government keeps assisting and helping Mexican people abroad that are affected by the current travel restrictions and makes a call for them to keep in touch with the network of embassies and consulates. (Logue, 2020)

The negotiation of the foreign ministry to bring these tourists back had to be done country by country given the different stops and migratory regulations each one adopted. The German government worked round the clock to deliver almost 40 airplanes to take home stranded German tourists, journalist Melinda Crane (2020) reports. Several EU countries followed different lockdowns with the same obstacles in which case negotiations had to be done slowly. The *New York Times* reporter Tariro Mzezewa (2020) alerts that the sentiment of abandonment interviewees often show is related to the poor information that they obtained regarding returning home. The governments said little about how and when these tourists would be repatriated. Thousands of Americans had been stranded in different tourist cities when COVID-19 surfaced. Mzezewa (2020) collected testimonies of different persons:

> "It's super frustrating not hearing anything from your government," Ms Eden said. "I feel completely abandoned."—Her sentiment was echoed by Moussa Diene, 44, an American citizen who said he is stuck in Casablanca and running low on his diabetes medication. Mr Diene, 44, said he has been to the consulate, but did not receive any help.

As a backdrop, the statement "suspend international flights until further notice" was commonly resonating in the media and digital platforms. People used blogs and platforms to coordinate food delivery and shelter and give humanitarian assistance to tourists. In Bauman's terms, the pandemic has blurred the borders between vagabonds and tourists who live now as real asylum-seekers. With the rules of free transit suspended, the capital was not enough to ask for hospitality: "Our government, embassy, and airlines don't seem to be helping us, so maybe we can help each other" is the description of the group's creator, Jennifer Bell, came up with" Mzezewa (2020).

CNN reporter Jessie Gretener (2020) overtly says thousands of Australians stuck abroad struggled, in several ways, to come back to Australia.

> The Department of Foreign Affairs and Trade (DFAT) says at least 25,000 Australians, many of who are financially and medically vulnerable, have registered their need to come home since July. However, the Board of Airline Representatives of Australia estimates the true number of those stranded is closer to 100,000.

Needless to say, Australia adopted one of the strictest lockdowns in the world, like New Zealand, leaving thousands of Australians stranded abroad. Some calculations reveal that more than 357,000 Australian citizens returned to their homes in the three months following the closure of the borders. As a consequence, many of these peoples lost their jobs (Gretener, 2020).

The Gaze of an Undesired Guest (Wicked Gaze)

In Chap. 6, we reviewed John Urry's works, in particular his conception of the tourist gaze (Urry, 2002). With the benefit of hindsight, he argues eloquently that the tourist gaze should be framed into a cultural matrix that molds our expectations, fears, and hopes about what is desirable, to gaze, or not. Having said this, the figure of the tourist occupies a central position in the social imaginary of global capitalism. Tourists, unlike migrants, are seen as the ambassadors of democracy and capitalist progress. They go through the safer paths of the empire while forced labor migrants have no such luck. We coin the term *wicked gaze* to describe the rise of a new narrative where tourists are not desired nor are they welcomed by the hosts. The term *wicked* equates with evildoer, someone who corrupts himself because of radical evilness. This does not say tourists are evildoers, but they are very well demonized given the potential danger (virus) they carry, especially when foreign tourists violated quarantine by leaving their hotel. On March 19, 2020, 1 Italian and 21 Korean citizens who visited Iguazu Falls in Argentina for tourism and leisure purposes were deported by immigration authorities after they were found to be violating lockdown restrictions. Monitored and detained by the Federal Airport Police after their refusal to be in isolation, the tourists were exiled. This happened mainly because, in the post-COVID-19 days, tourists were labeled as potential carriers (terrorists) disseminating a lethal virus. They, tourists, were the targets of aggression and different acts of violence and were carefully scrutinized as potential risks. In Argentina, the mass contagion started with the arrival of the first international flights that brought Argentinean expatriates, some of whom escaped from quarantine. In the days before the lockdown, interviewed front-line staff members were reluctant to make personal contact or to interact with foreign tourists coming from Europe or the US. As Virginia (24 years old, female) said:

> All of us are careful while attending foreign tourists, but I am particularly afraid to be in contact with American or European customers. Some of them can be infected and disseminating sickness without symptoms. It is very strange the situation, all our lives we have been educated to serve tourists, we speak foreign languages and are receptive to other cultures, but this virus upends everything, we are frightened about the situation if you ask me, I have never experienced anything like this before.

At first glance, Virginia is concerned about getting infected after a close interaction with a customer, but beyond her concerns lie many profound prejudices and ideological disposition oriented toward disparaging China and Chinese tourists. Roberto worked as a travel agent with a tour operator located in Buenos Aires. Per his viewpoint, "here the problem is communism and China, they live in totalitarian regimes which prohibits public acts or demonstrations. Some doctors were strictly censored by the communist government in 2019. The result seems to be clear, the sickness was widely disseminated through Chinese tourists worldwide. They are the real criminals who caused serious damage to the health system" (Roberto, 45 years old, male).

The consulted staff showed certain antipathy for foreign tourists, requesting the intervention of security forces to arrest and deport the infringers. Among the expatriates, many Argentinean citizens were living abroad. They were targets of chauvinist expression as well. Ariel, who is a hotel receptionist in Mendoza, stated:

> Our problem is a problem of ideology, the high-class prefer predominantly the international destinations, they have money to pay for that and fly to

> Europe or the North. Now, we still keep our airspace open, and many flights from Europe are coming. Not only we need to close the borders now, but leave Argentines abroad. I have never left my homeland, those who work abroad are betrayers, and now they want to come back? Why? They do not feel genuine Argentineans; they love Europeans and European lifestyles. They know about the pandemic when opted to travel abroad, and they do that because always is the same, rich people do not respect the rules. (Ariel, 28 years old, male)

Lastly, the world witnessed the precipitation of several ethnic or tribal disputes because of the sentiment of panic instilled by the virus. The "Other", the neighbor state, had to be kept away to stay safe. Stay at home was the only way to remain healthy and safe. In the same way, nations fell into a climate of anarchy and chaos where the health recommendations differed from country to country. Some leaders appealed for *herd immunity*, trying to preserve the local economy, while others imposed strict lockdowns; local authorities confronted central administration not only for the control of financial resources but also because of unresolved old political rivalries. What is more important, the world entered a stage of capitalism characterized by ongoing inter- and intra-state *geopolitical tensions.*

Geopolitical Tensions

The mistreatment of foreign tourists led to a hot debate about the coordination of new budget lines or bailout programs and ignited conflict and unparalleled geopolitical tensions. These tensions fueled the veins of the nation with the spirit of separatism and racism. For example, Dutch finance minister Jeroen Dijselbloem blamed the Southern European countries were a burden on the European budget. He said: "I can't spend all my money on liqueurs and women and then go ask for your support!" Prime Minister Hoekstra cast some doubts on the capacity of Spain to cope successfully with the COVID-19 situation because of the lack of budgetary capacity. In the same way, Donald Trump accused directly the Chinese communist government of obscuring or covering up vital information that would have helped contain the pandemic. He went to level of threatening the freezing of the funds of the WHO, which he called *a puppet of China.*

The relations between Argentina and Brazil were strained when Jair Bolsonaro alluded to Fernandez administration as a continuation of

Maduro's regime in Venezuela. In the past, Argentina received a continuous migration of workers from neighboring countries such as Bolivia, Chile, and Paraguay. Before the unilateral decision made by the government of Bolivia, Peru, and Chile to close the borders, part of the popular parlance blasted. Joanna, who serves as a housekeeping manager, claimed:

> Bolivia and Paraguay shut down their borders, they do not want my compatriots, some of them are stranded, other life there. Do you feel it is fair? They came here to my country and were welcomed, they lived from our hospitality, they fed our food, and drank our wine, they educated in our schools and were attended in our hospitals. I feel it is time to be harder with them, we need to re-found the nation and delineate carefully the migration policies for neighboring countries. Other partners said that we should deport all Bolivians and Paraguayans living here. (Joanna, 37 years old, female)

Similar partisan discourses could be seen in other interviews such as that with Vanessa, the housekeeper of a five-star hotel in Buenos Aires. Regarding the individualist (narcissist) behavior of tourists who violated isolation on their arrival and the businessmen who took advantage of the situation and fired workers or reduced wages, she stated:

> My salary was reduced by almost 50% because of COVID-19. After years of profits and economic prosperity for the hoteliers, it is very hard to think they have no money now. My boss snipped the government of Fernandez because he was a populist leader. He insisted on the urgency of dismantling the welfare state, strengthening the market and reducing substantially the state. What do they think now? The same, they are cynic, they bid for the government to pay the workers´ salaries while they amass their money. COVID-19 reveals our miseries and injustices. (Vanessa, 35 years old, female)

The financial support provided by the governments to mitigate the negative effects of the pandemic was not enough. Millions of people lost their jobs while thousands of firms went bankrupt. The industry faced one of the worst crises in its history since the inter-war period.

A Thanaptopic Society

Another interesting point to discuss is the use of digital technologies to make new forms of tourism. Maria (27 years old, female), a tourism bachelor, thinks

> we need to exploit the digital technologies to promote new forms of tourism, probably robots may attend guests at hotels, or in airplanes, but what is more important, we do not need to move to make tourism, there is a type of virtual tourism I have commented in furor in the North. Virtual tourism is the future, we need to put our efforts in that direction. The human contact, as well as the mobile lifestyle of Occident, is the key factor that propagates this pandemic.

Digital technology acts as a big brother, which has dual effects. On the one hand, it ensures more secure forms of relation, but at the same time, it develops a morbid taste for witnessing a type of new thanaptopic society. Given this, some interviewees confessed that they lived the situation like a horror show, a spectacle, or worse a game where the key players were us, those who avoided being infected and remained alive. Horacio is an expert tour agent of Ushuaia city in Tierra del Fuego Province. He accepts, "I see this like a big game, a theatre, or what is worse, a spectacle where mass-death is the big player. I sit down at home and watch TV; the pandemic and COVID are the topics of today, I'm a witness of how people are dying, and this is sorrowful, but inside me, it feels great because I am not sick" (Horacio, 49 years old, male). The same can be found in Virginia, when she says, "This is an invisible war against a virus, who like a terrorist looms to attack us; all of us will become sick of this decease, but only the strongest will survive. I am not saying that this is a punishment of God, but an opportunity to enhance our morality. This is like the movie *Hunger Games* where only the skilled warriors or the strongest participants will survive".

The same was repeated by Juan Manuel (male, 26 years old), a young flight attendant, who opined: "This is lived as a war in real-time, an apocalyptic show where all are struggling to survive! I walked by the streets the first days of the lockdown presidential decree and the streets were deserted, the roads without cars or trucks. As a surrealistic dream, it captivated very well my attention and scared me. Both ingredients of a big show I have never experienced in the past".

The idea of a virus which should be extirpated or the command (a bellicose jargon) to a new war against a virus was present in these interviews. The free circulation and transit that cemented the modern nation-state's architecture were mutating to decentralized forms, units, where local authorities had more legal weight than the central administration. In travel lines, international tourism was setting the pace for new local tourism where interstates travel bans triplicate.

The War Against a Virus

The metaphor of the war against a virus is replicated in almost all interviews. It varies from a strong discourse revolving around social Darwinism (e.g. in Virginia) and the aestheticization of horror (in Horacio). With the benefit of hindsight, the coronavirus disease is interpreted as a foundational event, a unique event where the future of tourism remains uncertain. Valeria (42 years old, female) is the CEO of an important rent-a-car company. She believes:

> The future of tourism is unclear, fuzzy if you want a foundational moment like 9/11. Things will surely change, for example in international travels, but what is worse, probably tourism as we know it will not exist any longer. All our jobs are at risk, and this is a point I do not want to think about. I try to be positive all days, day by day, think about what I can do now, not in the future. Things are out of my hands.

Moreover, another potential horizon of research indicates that COVID-19 seems to be the beginning of a new epoch where liberty, tourism, and democracy are in decline. Juan (25 years old, male) is a travel agent in Cordoba; he avows:

> This is a syndrome, invented by the Chinese or the Americans to dominate the world. This virus is not normal, it was created at a laboratory. As a global pandemic, of course, our means of transport disseminated the virus while our forms of production and consumptions were forced to a gridlock. We cannot live without the lockdown, but the same lockdown is exhausting the economy. If you ask me, the final is clear, like a tragedy; no matter the outcome, our current form of lifestyle will wither away forever. I should use a mouth-cover. I see at the streets people but I am unable to recognize who they are. All of us are cloned but separated by the sanitary barriers which are real borders! Without human interaction, we are not humans anymore.

Maria (27 years old, female), in the same vein, goes on to say:

> It is strange, indeed; a couple of years ago, the world was flattened, anyone had the right to travel elsewhere, not only moved by tourism but businesses,

> or sports. This was a world of high mobility and speed. Tourism was invoked not only by governments as a prosperous activity but defended as a right all human beings should access. Now, things have changed a lot; we live in a feudalized world where the barriers are posed by states which form the same nation; separatism and racism came to stay. Donald Trump accuses China … we need to learn of this new period. I do not know to what extent it is an opportunity, or a calamity. I think of my children and imagine how the next decades will be. Probably we will involuntarily leave a better or more sustainable environment. COVID-19 is creating changes, some of which are positive, others negative.

Last but not least, COVID-19 territorialized a new geography for not only the nations but also the contours of a new feudalized world. The lack of interest for the Other and the geopolitical tensions as well as the rise of long-dormant chauvinist discourses are a clear sign of the same phenomenon: the feudalization of the capitalist world, which prompted a radical change in the travel behavior as well as new cosmologies orchestrated to understand the neighbor. These new borders of tourism will be discussed in the following section.

The New Borders of Tourism?

In a fractured world where COVID-19 is radically shifting international tourism, it is safe to say the future of tourism coincides with a feudalized geography that poses serious challenges in the years to come. The relational contact, which is characteristic of leisure activities, seems to be banned. The figure of the foreign tourist is rejected and subject to hostile reactions by the interviewees. Signs of chauvinism, racism, and anti-tourist expressions mark the pace of a new epoch where the tourist becomes an *undesired guest*. As of December 20, 2020, the virus had infected 1,541,285, causing 41,813 deaths in Argentina. Despite the long lockdown which prompted an economic downturn, the government did not have sufficient PCR tests to administer, accompanied by the isolation of infected persons.

As a backdrop, Argentina is legally formed of 24 autonomous provinces which reserve their authority of organization of their territorial security and local justice, while remaining hand-tied to regulate customs and impose migratory traveling restrictions. The constitution prohibits the provinces from imposing travel bans or stopping free circulation. Apart from this, different provinces adopted countless measures oriented toward tightening border controls. Some of them, like Formosa, impeded the entrance of 7000 persons who were compelled to live in poverty and inhuman conditions for several months. Many citizens tried to cross rivers or rugged terrains. Journalists reported some fatal cases where people met their deaths while trying to cross the borders. The travel bans and restrictions, far from being standardized or regulated by the central administration, were unilaterally formulated and imposed by each province at its discretion. In the following months, some provinces urged for the opening of the frontiers to tourists but requested different permissions and requirements. What is more important, the new visitors were encouraged by local authorities to present health cards, assist cards with health coverage, and PCR testing accompanied by other bureaucratic documents and forms. Argentinean journalist Gabriela Origlia (2020) describes the situation of Solange Musse, a young girl who died waiting for her father to visit her in Alta Gracia, Cordoba. He began a long trip from Neuquen but was forced to come back after two tests gave dubious results. Instead of receiving support and advice from health authorities, Solange's father was banned from entering Cordoba. Meanwhile, Solange, after struggling against a strange cancer, finally died a couple of days later. Similar stories are heard everywhere. Cynthia Ruth (2020) eloquently evokes the story of Mauro Ledesma (male, 23 years old) who drowned while swimming across the Bermejo River to visit his daughter. Mauro was working in a milking parlor situated in Cordoba when Fernandez sanctioned the presidential decree. For the first seven months his wife Tamara (20 years old) received remittances from Mauro but things went off the rails when he was fired. Mauro decided to come back to Formosa with his family, Tamara and his daughter. On August 25, he proceeded to fill the corresponding form to receive legal authorization to cross Formosa's border but without any result. After several weeks of waiting, he decided to come back in a clandestine manner that will end his life. Gildo Insfran, who is the governor of Formosa, unilaterally imposed a strict lockdown that stranded almost 8000 people in the borders for several months. Martin Boerr (2020) from *La Nacion* wrote that local authorities estimated that almost 8300 were stranded in the borders between Formosa and Chaco, awaiting legal authorization for their entrance. The Supreme Court had ruled that Insfran allow the repatriation of these persons to their homes. Formosa had accepted only 6000 people, leaving others in the borders because of the local bureaucracy and because many citizens were unable to pay for their PCR testing, which cost ARS 5000 (Argentine pesos).

At the end of December 2020, Fernandez made it mandatory for each foreign tourist to take a PCR test on their arrival at Ezeiza Airport. At the same time, the flights from the UK were cancelled because of the mutation of a new virus strain. The EU and its 27 Union members agreed on coordinated efforts to constrain international flights from England. While France and England were working together to reach a deal to end the fluidity of persons through the Channel because of the new COVID-19 which disseminated 70% faster, several other countries imposed serious restrictions on the trade with the UK. Once again, Saudi Arabia, Kuwait, and Oman shut down their airports and borders to international passengers. Some of Denmark's authorities alerted that the new strain was found in Australia, Italy, and Netherlands (BBC, n.d., On 22 December 2020, https://www.bbc.com/news/world-europe-55404087).

The end of the year and the opening of the summer holidays led Fernandez to relax some restrictions. Not all destinations were open for travelling abroad. For example, Chile authorized some special fights while Uruguay kept the borders closed. Europe accepted flights coming from Argentina but made quarantine for 14 days mandatory. In the US, the restrictions varied in each state. While New York insisted on quarantine, the same was not needed in Florida. The foreign ministry speaks to us of low and high restrictions. Brazil and Mexico were destination of low restrictions, according to journalist Maria Nollman. Since December 2020, in Buenos Aires, all tourists who have travelled or remained in a city farther than 150 miles for more than 3 days should present a PCR test at their return. Last but not least, Brazil was in the epicenter of an international scandal when a local judge of Rio de Janeiro sanctioned the mandatory deportation of all foreign tourists because of the recent increase of COVID-19 cases in Buzios. Tourists were directed to leave the city within 72 hours, as per the decision.

Conclusion

This chapter gives some insights into the crisis accelerated by COVID-19 in the West after further review. We place the dilemma of the vampire into the foreground to explain the fears and anxieties revolving around the alterity in these dark moments. If the colonial expeditions degraded the so-called inferior "Other", it gives at least the opportunity to be re-educated. Gradually the Western expansion had serious costs for the capitalist system. The West forged *ethnocentric hospitality* to digest the "non-Western Other". In view of this, hospitality rested on a conditional criterion that dispossessed the alterity of its quintessential features (George, 2008). Molded to the European eye, leisure travels marked the beginning of a new global era. The attacks on the US soil in 2001 followed by the stock market crisis that occurred in 2008 broke the boom of mobilities, fracturing the illusion of a global world and accelerating the end of hospitality. Now, COVID-19, which is far from being a foundational event, gives to the West its coupe de grace. Lastly, we have described the ebbs and flows of COVID-19 into a model separated into five important points. The *undesired guest* is a radical transformation of the tourist gaze by the locals. The *geopolitical tensions*, among nations, are essentially characterized by uncoordinated efforts of nation-states to move in a climate of uncertainty and chaos and speak to us of a state of anarchy that marks the end of globalization. This process results in *a thanaptopic society* where the self maximizes its pleasure through the "Other's death". The risk society imagined by Beck is over-written by a new dark society where the main commodity of exchange is death. Such a process is possible through the digital technologies that scrutinize our daily lives. *The war against a virus* inscribes as a powerful metaphor to keep on the securitization discourse initiated after 9/11. The war on terror sets the pace for a new war against a virus. All these points converge with the structuration of an emerging *feudal world* where the national and domestic borders are in constant change. Digital technology does not offer an opportunity to the tourism industry and leisure travels; rather, the virtualization of tourism practices evinces the crisis of the industry as well as the radical shifts of a world (new normality) without tourists.

References

Aldiss, B. (1997). Foreword: Vampires—The ancient fear. In J. Gordon & V. Hollinger (Eds.), *Blood read: The vampire as the metaphor in contemporary culture* (pp. ix–xiii). University of Pennsylvania Press.

Auerbach, N. (1995). *Our vampires, ourselves.* University of Chicago Press.

Avraham, E. (2020). From 9/11 through Katrina to Covid-19: Crisis recovery campaigns for American destinations. *Current Issues in Tourism*, 1–15. https://doi.org/10.1080/13683500.2020.1849052

Baudrillard, J. (1994). *Simulacra and simulation.* University of Michigan Press.

BBC. (n.d., December 22). Coronavirus. EU tries to agree response to new UK strain. *BBC.* https://www.bbc.com/news/world-europe-55404087

Boerr, M. (2020, November 22). unos 8000 varados comienzan a entrar a Formosa, pero le piden tres hisopados. https://www.lanacion.com.ar/politica/unos-8000-varados-comienzan-ingresar-formosa-pero-nid2517752

Cansola, A. (2020, May 6). Covid 19: Mexico to repatriate thousands of tourists stranded abroad. https://www.eluniversal.com.mx/english/covid-19-mexico-repatriate-thousands-tourists-stranded-abroad

Crane M. (2020, April 15). Coronavirus: German government to fly home tens of thousands of tourists. *DW*. https://www.dw.com/en/coronavirus-german-government-to-fly-home-tens-of-thousands-of-tourists/a-52811964

Derrida, J., & Dufourmantelle, A. (2000). *Of hospitality*. Stanford University Press.

Filimonau, V., Derqui, B., & Matute, J. (2020). The COVID-19 pandemic and organisational commitment of senior hotel managers. *International Journal of Hospitality Management, 91*, 102659.

Franklin, A. (2007). The problem with tourism theory. In I. Altejevic & A. Pritchard (Eds.), *The critical turn in tourism studies: Innovative research methodologies* (pp. 131–148). Elsevier.

George, B. P. (2008). Local community's support for post-tsunami recovery efforts in an agrarian village and a tourist destination: A comparative analysis. *Community Development Journal, 43*(4), 444–458.

Gordon, J., & Hollinger, V. (1997). Introduction: The shape of vampires. In J. Gordon & V. Hollinger (Eds.), *Blood read: The vampire as the metaphor in contemporary culture* (pp. 1–7). University of Pennsylvania Press.

Gössling, S., Scott, D., & Hall, C. M. (2020). Pandemics, tourism and global change: A rapid assessment of COVID-19. *Journal of Sustainable Tourism, 29*(1), 1–20.

Gretener J. (2020, September 14). They have forgotten us: Thousands of Australians stranded overseas in the face of governments stringent border controls. https://edition.cnn.com/travel/article/australians-stuck-abroad-covid/index.html

Hao, F., Xiao, Q., & Chon, K. (2020). COVID-19 and China's hotel industry: Impacts, a disaster management framework, and post-pandemic agenda. *International Journal of Hospitality Management, 90*, 102636.

Higgins-Desbiolles, F. (2020). Socialising tourism for social and ecological justice after COVID-19. *Tourism Geographies*, 1–14. https://doi.org/10.1080/14616688.2020.1757748

Howie, L. (2012). Witnessing terrorism. In *Witnesses to terror* (pp. 155–175). Palgrave Macmillan.

Ioannides, D., & Gyimothy, Z. (2020). The COVID19 crisis as an opportunity for escaping the unsustainable global tourism path. *Tourism Geographies*. https://doi.org/10.1080/14616688.2020.1763445

Ivanova, M., Ivanov, I. K., & Ivanov, S. (2020). Travel behaviour after the pandemic: The case of Bulgaria. *Anatolia, 32*(1), 1–11.

Korstanje, M. (2016). *The rise of thanacapitalism and tourism.* Routledge.
Korstanje, M. (2017). *Terrorism, tourism and the end of hospitality in the West.* Palgrave Macmillan.
Korstanje, M. E. (2019). *Terrorism, technology and apocalyptic futures.* Palgrave Macmillan.
Korstanje, M. E. (2020). Passage from the Tourist Gaze to the Wicked Gaze: A case study on COVID-19 with special reference to Argentina. In *International case studies in the management of disasters.* Emerald Publishing Limited.
Korstanje, M. E., & Olsen, D. H. (2011). The discourse of risk in horror movies post 9/11: Hospitality and hostility in perspective. *International Journal of Tourism Anthropology, 1*(3–4), 304–317.
Korstanje, M. E., & Tarlow, P. (2012). Being lost: Tourism, risk and vulnerability in the post-'9/11' entertainment industry. *Journal of Tourism and Cultural Change, 10*(1), 22–33.
Logue, J. (2020, May 14). We are abandoned: These tourists stranded in the US because the coronavirus want to go home. *ABC NEWS.* https://www.nbcnews.com/news/latino/we-are-abandoned-these-tourists-stranded-u-s-because-coronavirus-n1206386
Mzezewa, T. (2020, March 18). American stranded abroad: I feel completely abandoned. *New York Times.* https://www.nytimes.com/2020/03/18/travel/coronavirus-americans-stranded.html
Nollman, M. (2020, November 12). Vacaciones y COVID19: que se les pide a los argentinos que viajan al exterior. *La Nacion.* https://www.lanacion.com.ar/sociedad/que-destinos-les-exigen-cuarentena-argentinos-ingresar-nid2466889
Origlia, G. (2020, August 25). ¿Aduanas provinciales? Con hisopado y burocracia restringen el libre transito. *La Nacion.* https://www.lanacion.com.ar/politica/aduanas-provinciales-con-hisopados-burocracia-restringen-libre-nid2431008
Pearce, P. L. (2019). Are tourists interesting? In P. Pearce (Ed.), *Tourist behaviour* (pp. 2–19). Edward Elgar Publishing.
Polyzos, S., Samitas, A., & Spyridou, A. E. (2020). Tourism demand and the COVID-19 pandemic: An LSTM approach. *Tourism Recreation Research, 44*(2), 175–187.
Prayag, G. (2020). Time for reset? COVID-19 and tourism resilience. *Tourism Review International, 24*(2–3), 179–184.
Radic, A., Law, R., Lück, M., Kang, H., Ariza-Montes, A., Arjona-Fuentes, J. M., & Han, H. (2020). Apocalypse now or overreaction to coronavirus: The global cruise tourism industry crisis. *Sustainability, 12*(17), 6968–6970.
Rice, A. (1976). *The vampire chronicles.* Knopf.
Rives-East, D. (2019). *Surveillance and terror in post-9/11 British and American television.* Palgrave Macmillan.

Rogerson, C. M., & Rogerson, J. M. (2020). COVID-19 tourism impacts in South Africa: Government and industry responses. *GeoJournal of Tourism and Geosites, 31*(3), 1083–1091.

Seyitoğlu, F., & Ivanov, S. (2020). Service robots as a tool for physical distancing in tourism. *Current Issues in Tourism, 24*(12), 1631–1634.

The Hunger Games. (2012). Gary Ross (dir.). Lionsgate Films, 142 Minutes, the US.

The Lost Boys. (1987). Shumacher J (dir.). Warner Bros. 98 Minutes, the US.

Tzanelli, R. (2016). *Thanatourism and cinematic representations of risk: Screening the end of tourism*. Routledge.

Tzanelli, R. (2018). Schematising hospitality: Ai WeiWei's activist artwork as a form of dark travel. *Mobilities, 13*(4), 520–534.

Urry, J. (2002). *The tourist gaze*. Sage.

Wen, J., Wang, W., Kozak, M., Liu, X., & Hou, H. (2020). Many brains are better than one: The importance of interdisciplinary studies on COVID-19 in and beyond tourism. *Tourism Recreation Research, 46*(2), 310–313.

Zanger, J. (1997). Metaphor into metonymy: The vampire next door. In J. Gordon & V. Hollinger (Eds.), *Blood read: The vampire as metaphor in contemporary culture* (pp. 17–26). University of Pennsylvania Press.

CHAPTER 9

Conclusion

When we launched this book project, COVID-19 was already paralyzing global trade and placing governments on the brink of collapse. For some, this event aligned with an apocalyptic moment of humankind that evoked self-introspection and the urgency of a rebirth. This book instead goes in the opposite direction. The current crisis of identity started with the terrorist attacks of 2001 on US soil. We preferably center on Western civilization's problems to understand and practice hospitality, at least giving shelter and food to undesired guests. In several parts of the book, we use the term *undesired guest* to mean those human agents who are often not welcomed in a country. The nation-state has evolved according to its notion of hospitableness. To be open to the stranger is not only a moral duty but is also a sign of superior evolution over other cultures.

This doctrine has involuntarily created serious points and disputes over recent years because of the rise of mass migration. Thousands of asylum-seekers and displaced people are now knocking on the doors of heaven, which, so to speak, remain closed to all of them. Writing a book about hospitality seems not to be an easy task to perform, but we have the opposite figure: the tourist. The question of whether migrants are vagabonds—citing Bauman—who ask for absolute hospitality tourists have desired guests who are invited to travel, consume, and gaze the entire world. As Bauman puts it, the right of traveling for one does not correlate with the lack of mobilities of the others. These asymmetries, which were

M. E. Korstanje, B. George, *Mobility and Globalization in the Aftermath of COVID-19*,
https://doi.org/10.1007/978-3-030-78845-2_9

characteristic of capitalism, were aggravated after 9/11 and the "war on terror" declared by the Bush administration. Terrorism not only instilled a climate of paranoia and fear but closed the doors of Eden, probably without a real sin (Bauman, 2017). Meanwhile, we lay our fingers on the fact that the Western civilization, in its history, amounted in accordance with an ethnocentric view (paternalist) of the "non-Western Other". As discussed, the "Other" represented an inferior child, like a being that needed to be re-educated according to the cultural values of the Occident. The Western expansion cultivated the interests for leisure travels, literature, and science and engendered a sentiment of the culprit that was sublimated through captive writings. As a new literary genre, captive writings told the stories of hapless European travelers who were attacked and abducted by savages.

Torn between two civilizations, captives were portrayed as real heroes who sacrificed their well-being for the glory of civilization. Captive narratives fed the curiosity for the *non-Western Other* while captivating the attention of a vast readership. In the following centuries, this genre set the pace for a set of novels oriented toward narrating the obstacles and problems of scientists in their voyages. The imperial voyages, which inspired the rise of the first ethnographies in overseas territories, gradually crystallized in the travel literature, a new descriptive genre that punctuated the lifestyle in the overseas colonies as well as the life in tribal communities. This literary genre included journalist and documentary reports or ethnographers' fieldwork notes. The process of decolonization accelerated by the end of World War II opened the doors to new narratives to keep the former colonial periphery under control. The development theory and the ideals of democracy and free trade were the bulwarks of the West, now toward a new stage of economic maturation. It is important not to lose sight that thousands of migrants coming from the former colonies populated the growing European cities, forming stable manpower that consolidated global capitalism. Once the West's external borders were blurred, the doctrine of the enemy looming from the outside became a problem very hard to grasp. Paradoxically, globalization was enthusiastically embraced as the triumph of liberalism and national liberalism over other ideologies, but in so doing, these non-Western "Others" now living with us were packaged as the new internal enemy. Of course, the doctrine of the enemy living within resonates with force after the attacks of 2001 in the US.

While the Western civilization cemented technological advances that gave economic prosperity, stability, and peace as never before, Westerners feel more insecure now than in earlier decades. The turn of the century witnessed an increase in the risks and dangers that place civilization's integrity in jeopardy. Real or not, these global threats interrogate further the ways the "non-Western Other" is treated. The liberal market recreated the conditions to exploit a hypothetical future that never takes place in reality. Doubtless, modern terrorism ignited a social decomposition that affected social ties. The West not only closed its borders to strangers but also the "Other" became an *undesired guest*. Lastly, the recent virus outbreaks known as SARSCOv2 whips today the global trade, but, far from being a foundational event or a moment of reflection and rebirth, it affirms a process started in 2001. In this respect, COVID-19 should be understood as the foreclosure of an epoch, when modern tourist travels changed forever. We cannot conclude a book dedicated to the sociology of travels and geopolitical relations without discussing the nature of the passport. Although much has been said regarding the sociology of travels, less is known about the *passport*. In times of COVID, some voices have emphasized the urgent need to create and deliver an *immunity passport* to distinguish between healthy and already-infected people. The issue occupies a central position in this concluding chapter as the gateway toward new mobilities. What is a passport?

Since time immemorial, travelers had to produce different documents and letters of recommendation to travel. Etymologically speaking, the term *passport* comes from *a medieval document* mostly presented to pass through a medieval city gate (*porte*). These documents were issued by local authorities to testify to the intentions and identity of the beholders. These travelers were in transit in a set of cities previously declared in the document. Although Henry V of England is credited to be the inventor of the passport, which helped travelers to prove their identities abroad, it is also true that the history of the passport traces back to the twentieth century when it was used to regulate the mass migration to the US after the end of World War I. Between 1921 and 1924 the US passed two important laws known as *the Emergency Quota Act* and the *Immigration Act*. The original intention was to classify the origin of these newcomers and identify some disruptive undesired elements such as anarchists and potential terrorists. The passport served as a mechanism of control. The function of the passport was double-fold. On the one hand, it protected those strangers who asked for hospitality, such as migrants, asylum-seekers, and political exiles (to name a few). On the other hand, it marked and disciplined the newcomers classifying them according to the interests of the nation-state. As Martin Lloyd (2003) puts it, in ancient times, passports were documents issued to protect enemies' identity during negotiations or after the celebration of certain peace treaties.

It is important to add that in earlier works we argued that modern industrial tourism is terrorism by other means. We are educated to believe terrorism is an illegal act of political violence against innocent people, while tourism can be understood as a leisure practice derived from the modern industrial revolution which includes paid holidays and working hour reductions. Both things are true, but historians have misjudged the intersection between terrorism and tourism as well as their commonalities. In recent decades, especially after 9/11, it is not surprising to see how modern jihadists have directly targeted international tourists or bombed leisure spots and tourist destinations worldwide (Korstanje & Clayton, 2012; Korstanje, 2018). Doubtless, the origins of modern terrorism and tourism have the same roots. With the advent of the industrial revolution, passports were issued to monitor overseas migrants. In this respect, the industrial revolution gravitated seriously in local economies, pushing many farmers to migrate to cities and European urban geographies. Thousands of Europeans impoverished by the era of capital opted to migrate to the US, Argentina, or Australia.

In a seminal book entitled *The Anarchist*, historian James Joll describes the indoctrination processes of the European migrants in the US, a new land where the industrial revolution was in its infancy. Many of these newcomers were recruited to work in factories but subjected to inhuman conditions of labor. Some anarchists who came from Italy and Germany decided to fight against the nation-state, planning attacks against police officers and politicians. Labeled as terrorists, they were trapped, tortured, and exiled to Europe. Nevertheless, a few of them devoted their time and efforts toward organizing worker unions in the US. A new movement known as *anarcho-syndicalism* pushed to achieve better labor conditions such as paid holidays, reduction of working hours to eight, and the right to strike and thus laid the foundations toward unionism (Joll, 1964). The nation-state, through selection, stripped anarchism of its negative effects, adopting its ideological core as the touchstone of the capitalist system. In the process, the invention of the modern passport was vital to select the desired from the undesired "Others". In his book *The Society Must Be Defended*, Michael Foucault (Foucault & Ewald, 2003) gives a fresh insight in this direction. He argues eloquently that the mechanism of discipline consists of placating the destructive features of hazards and converting them into acceptable risks. He equates society to a living organism, threatened by external dangers (like viruses). The vaccine is a mitigated virus stripped out of its destructive effects. In the same way, the discipline adopts the core of external dangers, placating their devastating consequences. The same metaphor very well applies to the connection between terrorism and modern tourism. While the first anarchists (terrorists) were extirpated, their ideological core served the system to promote revolutionary labor advances that crystallized in expanding modern capitalism. Worker unions achieved important labor benefits and advantages which encouraged the evolution of leisure travels and the modern tourism industry. The invention of leisure travels sedated the hostilities of anarcho-terrorist cells while affirming the ruling elite's power.

The figure of the passport played a leading role in distinguishing the desired from the undesired guests, separating wheat from the chaff. To some extent, the passport acts as an instrument of discipline which selectively gives certain order to the chaotic presence of the "Otherness". According to migratory requirements, each passport follows a nationality that receives different treatments by the nation-state. Today, people migrating from countries suspected of terrorist activities are tightly scrutinized by the migration officers before entering the US while tourists coming from Anglophone countries are not disturbed. The passport speaks to us of a hierarchy of nationalities where some have few travel restrictions. Having said this, the history of passport is inextricably intertwined with leisure travels. In *The Passport in America*, Craig Robertson (2010) indicates that the origins of passports in America are associated with the need to control those newcomers who are poor, sick, or potential criminals without mentioning their insufficiencies. Passport communicates something important about the holder's identity and allows a more accessible and more efficient classification. Basically, the passport and visa are given to answer the basic questions enumerated by Derrida: *who are you and what do you want?* But the history of the twentieth century is accompanied by a technological revolution that cemented the authority of the nation-state as the most important geopolitical entity. To put it differently, the history of passport inscribes in the evolution of the national identity, which is conducive to the expansion of travels and the needs of dealing with the undesired "Other". Robertson (2010: 12) goes on to write:

> Identification in the US border made use of dominant understandings of the body, personal appearance, and identity that were supported by scientific and popular claims for the reliability of exterior signs to ascertain an individual's true, essentially identity; depending on the circumstances, that identity could be an individual's race, class, gender or character. Immigration policy meant that officials only needed to become astute observers of prostitutes, imbeciles, and potentially unproductive immigrants, either diseased or lazy.

The current uses of travel visas seem to be pretty different from the practices or requirements at the beginning of the twentieth century. Now, travelers who visit neighboring countries need neither visa nor passport, but a passport is indispensable if the travel takes one to other continents. As discussed, the passport marks the gateway for the passenger credits, not only its identity but to give the authorization to move on. Those who lack a passport are limited to domestic travel.

What does the COVID-19 immunity passport mean? The immunity passport is a new term coined by experts during the global pandemic of COVID-19. It corresponds with the issuance of a certificate that ascertains that some individuals have sufficient immune antibodies to SARSCOv2. Originally it endorsed further liberty to the beholder to travel and cross the borders. This activated the tourism industry, allowing temporary entrance to some tourists while restricting others who have not been infected. The *immunity passport* replaces the modern logic of the classic passport inserted in the national identity by the health condition. Here two assumptions should be made. On the one hand, the document associates with other requirements such as safe destinations and risk-free certificates. One might speculate those who have been infected or vaccinated will be the new privileged tourists legally allowed to move. The old dichotomies of tourists versus vagabonds placed by Bauman set the pace for a new reality, where healthy persons are immobilized. On the other, since the travel behavior changes, it does not surprise that the travel restrictions change as well.

This book interrogated further the origin, expansion, and evolution of Western travel since the colonial epoch to date. The Western civilization, since 9/11, seems to be experiencing a fear of the stranger, which has led to the decline of hospitality. Of course, the future of the tourism industry and mobilities remains uncertain. We contend that, far from being a turning point, the COVID-19 reinforces a previously established spirit aligned with the "non-Western Other" demonization.

In Chaps. 1 and 2, we discussed the orientation of the Western narrative to stimulate overseas expeditions and travels beyond the comfort of security. These narratives institutionalized long-distance navigation and convinced monarchies to commit significant resources for furthering the infrastructure needed for its continued development. More roads and ships were built, and new navigation routes were charted. These travelers laid the foundation for a new Westernized epistemology of mobility and created an ethnocentric view of hospitality, respecting the strangeness. *There is supposed superiority for those who arrive over those who are arrived.*

Chapters 3 and 4 dissected the invention of the liberal market and free trade as a dominant narrative of liberal thinking to govern daily life. As stated, we cannot imagine the capitalist society without the invention of risk and subsequent solutions to contain it. But these conventions are no real entities but social constructs which never take place as they are imagined. This invariably leads us to decipher the philosophical dilemmas beyond John Urry's paradigm and his conception of the tourist gaze. He formulated a convincing model applicable to a hyper-mobile world where tourists were ambassadors of prosperous nations. This conception has radically shifted because of the recent outbreak of COVID-19 and the newest strain mutation. If anything, tourists from the US are now detested more than ever for the fear that they carry and spread the virus. The US's extremely poor performance in fighting against the virus meant the wealth of tourists does not automatically mean them being welcomed.

Complementarily, Chaps. 5 and 6 investigate the commonalities between terrorism (with emphasis on the lone wolf archetype) and COVID-19. Using bellicose jargon, in both cases, humanity declares war on an inhuman entity. The war on terror relates to human emotion, while the war against COVID to an actual virus. It sounded as if the Americans were more united in the former war. Or, did the "live with it" narrative that they heard about the former seep into their response to the latter? Hard to say, and more investigation is needed. In the days after 9/11, anyone who looked and talked differently would be a potential terrorist; now, this includes even those who are more or less alike, although Asians and Chinese Americans bear most of the brunt of it. The problems of the new travel behavior, the re-drawing of human ties and global geographies, and the long-lasting effects of terrorism in the secrecy plots, as a continuation of captive novels, are carefully explored in Chaps. 7 and 8.

Finally, a short conclusion closes the book that, at least for the authors, brings some critical reflection on and light to the dark days that we are living in.

References

Bauman, Z. (2017). Tourists and vagabonds: Or, living in postmodern times. In *Identity and social change* (pp. 19–32). Routledge.

Foucault, M., & Ewald, F. (2003). *'Society must be defended': Lectures at the Collège de France, 1975–1976* (Vol. 1). Macmillan.

Joll, J. (1964). *The anarchists.* Eyre & Spottiswoode, p. 121.

Korstanje, M. E. (2018). *Terrorism, tourism and the end of hospitality in the West.* Palgrave Macmillan.

Korstanje, M. E., & Clayton, A. (2012). Tourism and terrorism: Conflicts and commonalities. *Worldwide Hospitality and Tourism Themes, 4*(1), 8–25.

Lloyd, M. (2003). *The passport: The history of man's most travelled document.* Sutton Publishing.

Robertson, C. (2010). *The passport in America: The history of a document.* Oxford University Press.

Epilogue: New Travel Behavior in the Era of COVID-19

At the end of December 2019, a new virus outbreak originated in Wuhan, China, and paralyzed the tourism industry and the global trade. SARSCOv2, known as COVID-19, has led the industry to an unprecedented crisis. Within months the virus propagated to five continents and has situated as a global pandemic as well as one of the nightmares of Western civilization. Because of this, scholars moved to analyze the future of tourism and debated different steps and protocols to mitigate the negative effects of the pandemic. In the incipient literature, theories revolving around COVID-19 and tourism are mainly organized into two clear-cut poles: those studies that are interested in developing new methodological lines to overcome the current limitations of tourist-centricity, which means the obsession for tourists as the only reliable actors of the system, and those focusing on social distancing and degrowth opportunities to reverse the current levels of contamination. Despite their differences, both academic waves agree that COVID-19 was a disrupting but foundational event that radically changed Western lifestyle and travel behavior.

In earlier works, we, rather, hold the thesis that, far from being a foundational event, the coronavirus disease affirms a previous trend that originated in 9/11 and the war on terror declared by the Bush administration. Dubbing this epoch as "the death of hospitality", we certainly found some commonalities between COVID-19 and 9/11 (beyond the strange

M. E. Korstanje, B. George, *Mobility and Globalization in the Aftermath of COVID-19*,
https://doi.org/10.1007/978-3-030-78845-2

repetition of the numbers nine and one). To some extent, terrorism ignited a new face of Western lifestyle where "the Other" was the subject of fear and mistrust. If the colonial travels were inspired in need of meeting with a radically different "non-Western Other", the culture of fear, which is shaped by terrorism, depicts the neighbor, who looks like us, as a potential terrorist. In the days of COVID-19, all citizens are potential terrorists (carriers) who threaten the social order. The video game *Among Us*, -which has captivated young gamers, serves as a clear example punctuating the idea of living with the enemy inside.

Secondly, and most importantly, digital surveillance technologies structured new forms of control at borders and airports, which further interrogate travelers. As Derrida brilliantly observes, hospitality can be offered or not, but whatever the case may be, two questions are vital when we meet a stranger: who are you and what do you want? The figure of visa occupies a central position in the ritual of hospitality. At a closer look, it can be seen that the word visa stems from *visum*, a Latin verbal past tense associated with *videre*, which means "to see". The hosting state, echoing our French philosopher, needs to check our identity before proceeding to offer hospitality. To the traditional documents requested of travelers such as a round-trip ticket and a visa or passport, we must add the COVID-19 test and temperature test. The history of passport speaks to us of the outsider's problem and homeland security. The passport was a medieval document issued to travelers who needed to move from one city to another. The document signaled the beholder's identity as well as the entrance gate number. Any outsider has temporal permission to stay at the city, under the auspices of the king, for businesses, leisure, trading, or any other purpose. Etymologically speaking, the term *passport* stems from the Latin words *pass* (crisscrossing) and *porte* (gate). King Henry V of England was the pioneer in coding a system of the modern passport to identify and classify outsiders. Through a novel passport system, these unknown visitors proved their real identity. Of course, the rite of hospitality closes the tension generated in host-guest encounters (Korstanje & George, 2021). As Derrida puts it, hospitality might be offered or not, but in any case, two questions remain: who are you and what do you want? (Derrida & Dufourmantelle, 2000). It is safe to say that either identity occupies a central position in the configuration of the rite of hospitality or destination security. More importantly, in the global world, passports take part in a hierarchal system of nationalities where some have accesses to multiple destinations while others are seriously restricted or limited to present further documentation. In the days before the pandemic, the world was divided into stronger and weaker passports. Nowadays, travel bans and restrictions that are being imposed on foreign and domestic tourists have ground the tourism industry to a halt, transforming not only our travel behavior—without mentioning the global geography—but also the nature of the passport. Besides the question of identity, the passport of the future will be designed to distinguish a safe (sane) visitor from a potentially infected one. The health (vaccine) passport poses a serious philosophical dilemma for those who refuse to be vaccinated. If persons who reject the vaccine are not endorsed to a health passport, then they are unable (hand-tied) to move. In this respect, the universal right of traveling, at least as it was discussed in earlier works (see Bianchi et al., 2020; Baum & Hai, 2020), is being transformed according to the levels of immunity of tourists, marking the start of a new mobility: pandemic mobilities. Last but not least, the passport of the global era tested the credibility of visitors as well as their identity and travel purposes while the COVID-19 passport scrutinizes closely human biology (knowing whether the tourist is infected or not). Although these passports are widely supported by digital technologies and platforms creating a revolution in the tourism industry, no less true is that in this new feudalized world, the option of lying about our real identity (a point very well noted in migrants' story lives) is no longer possible. Identity is culturally negotiated and cemented whereas immunity is biologically determined. Identity can be falsified or fabricated but not immunity. Besides, those tourists coming from richer countries who have access to the vaccine will start to travel with health passports while other citizens who dwell in poor societies will be left in complete immobility. In the past, capital marked the difference between tourists and vagabonds, to cite citing Bauman, but now sanitary reasoning is taking the lead in differentiating *desired* from *undesired guests* (Korstanje & George, 2021). Needless to say that this tendency (of anti-hospitality) started with the war on terror declared by George W. Bush.

Metaphorically speaking, in the war on terrorism technology scrutinized the travelers' luggage suspecting a bomb inside; now the threat is supposed to be carried by the traveling guest. The war on terror set the pace for the war against a virus. If technocracy reminds the ideological system of governance based on technology to frame the decision-making process, no less true is that "tourist travels" are shifting to "technocratic travels". In post-pandemic tourism, the "Other" is an undesired guest.

Third, both terrorism and COVID-19 operate in the context of high uncertainty, which is characterized by the rise of "a feudalization process".

In some words, let's describe to readers the situation in Argentina these days. In March 2020, president Alberto Fernandez implemented countless measures to stop the contagion, from the closure of airspace, borders, restaurants, and factories to the cancellation of public spectacles, events, and commercial activities (to name a few). In this respect, thousands of Argentinian tourists were suddenly stranded abroad. The foreign ministry played a leading role in the coordination of repatriations. At the same time, many foreign tourists wandered on the streets of Buenos Aires, looking for a way to be repatriated to their homes. Something had to be changed. In the following days, police and migration officers patrolled the streets looking for foreign tourists who had violated the lockdown. They were systematically traced, put on trial, and exiled. All visitors from neighboring countries such as Brazil or Chile were required to be isolated for 14 days. The archetype of the tourist, who had been historically seen as an ambassador of civilization, has now been re-typified as an ambassador of disaster. Some estimations reveal today that almost 4000 people were stranded between the provinces of Chaco and Formosa in November 2020. The lack of standardized methods to regulate mobilities in Argentina led the provinces to adopt different measures. John Urry eloquently imagined the figure of the tourist gaze to denote a much deeper cultural matrix which is finely enrooted in global capitalism, but now this tourist gaze mutated to a wicked gaze. The "Other" who is seen as a potential danger should be strictly monitored and controlled, if not eradicated, to preserve order. It is safe to say that COVID-19 seems not to be the cause but a symptom of a moral decomposition, a feudalized world where the "Other" is negated. We are thinking in the metaphor of the vampire who needs an invitation to step in, like Dracula, the vampire—as the supreme evildoer—needs to be invited before he can drink the blood of his victim. Likewise, the demon needs an invitation to possess his victim too. This suggests that the nature of evilness is given to those who violate or are unable to offer hospitality to their guest. Without any doubt, the lockdown accelerates some material asymmetries as well as the suspension of the right of traveling, at least through the territory. In the feudalized Argentina, provinces closed their borders and probably tightened the controls and migratory documents. The main contention of this short note is that we are living a feudalizing process vividly marked by a strong fear for the stranger. Scholars should pay attention to the sociological effects of COVID-19 on culture in the years to come. What is equally important, some provinces attune efforts to boost tourism to revitalize their economies, opening their internal frontiers. While visitors should present the respective COVID test document certifying they are healthy, other requirements such as an all-inclusive tour or round-trip tickets without mentioning certificate of residence or health insurance for travels are strictly requested upon arrival. All these noted measures are requisitions for traveling abroad but hitherto not applied for domestic tourism in normal conditions.

Last but not least, we have discussed how global geography is undergoing radical transformations. The question of hospitality is of paramount importance to understand the radicalized discourses of hate, against foreigners, and the social decomposition of the current epoch. Islamophobia, tourist-phobia, Windrush scandal, Brexit, hostilities against Asian tourists who are suspected of disseminating the virus, the Wall in Mexico-US border, and now social distancing are clear symptoms of the end of hospitality, at least as we know it. The (non-Western) "Other" has become an undesired guest!

In some respects, terrorism and the attacks on the World Trade Center remind two important things. On the one hand, a radicalized group (Al-Qaeda) humiliated the most powerful country in the world, employing four civilian airplanes which, so to speak, were valorized as the hallmark of modern mobilities. From that moment on, lay-citizens asked for tighter controls at the borders and abandoned public life. Living at home, or stay at home, is considered the best option in an ever-changing world. On the other hand, the securitization process introduced after 9/11 speaks to us of a new climate of fear, characterized by Jean Baudrillard as "the great simulacra". In a futurist world organized to achieve a zero-risk society, technology governs our bodies from the future imposing hypothetical events which never take place as imagined. Home exhibits a fertile ground to re-connect inert bodies in the plug of a pseudo-reality. The curiosity for the Other's death and the obsession for consuming disasters, in a thanaptopic society, are signs of the same social decomposition we dubbed as "the end of hospitality".

References

Baudrillard, J. (1994). *Simulacra and simulation*. University of Michigan Press.

Baum, T., & Hai, N. T. T. (2020). Hospitality, tourism, human rights and the impact of COVID-19. *International Journal of Contemporary Hospitality Management, 32*(7), 2397–2407.

Bianchi, R. V., Stephenson, M. L., & Hannam, K. (2020). The contradictory politics of the right to travel: Mobilities, borders & tourism. *Mobilities, 15*(2), 290–306.

Derrida, J., & Dufourmantelle, A. (2000). *Of hospitality*. Stanford University Press.

Korstanje, M., & George, B. (2021). *The nature and future of tourism: A post COVID19 context*. Apple Academic Press.

Urry, J. (2002). *The tourist gaze*. Sage.

Index

M. E. Korstanje, B. George, *Mobility and Globalisation in the Aftermath of COVID-19*,
https://doi.org/10.1007/978-3-030-78845-2

The manufacturer's authorised representative in the EU is Springer Nature Customer Service Centre GmbH, Europaplatz 3, 69115 Heidelberg, Germany. If you have any concerns regarding our products, please contact ProductSafety@springernature.com

Printed and bound by CPI Group (UK) Ltd, Croydon, CR0 4YY
30/09/2026
02225455-0001